高等职业教育**餐饮类专业**系列教材

咖　啡

（双语）

主　　编：钟小立　陈丽丽　母　海
副 主 编：黎梓亮　应灵弈　陈法拉
参编人员：叶宁青　韩梦蕾　吴　璇

重庆大学出版社

内容提要

为进一步提升餐饮、旅游、酒店类专业学生和从业人员的国际服务职业素养、职业道德、职业技能和可持续发展能力，本书基于行业的职业需求，结合项目教学与任务教学的形式，将咖啡服务和英语应用融合在一起，既突出咖啡专业知识与技能，又强化英语实践技能训练。《咖啡（双语）》选材新颖，内容丰富，以中英双语介绍咖啡基础知识、咖啡加工、咖啡冲泡、咖啡拉花艺术和咖啡馆文化5个项目。本书是一本注重语言灵活运用、人文素养和职业综合能力深度融合的行业英语教材。本书可以作为职业院校餐饮类专业教材，也可以作为社会学习者在职业生涯规划中提升职场能力和专业竞争力的自主学习资料。

图书在版编目（CIP）数据

咖啡：汉文、英文 / 钟小立，陈丽丽，母海主编
. -- 重庆：重庆大学出版社，2024.5
高等职业教育餐饮类专业系列教材
ISBN 978-7-5689-4331-4

Ⅰ.①咖⋯ Ⅱ.①钟⋯ ②陈⋯ ③母⋯ Ⅲ.①咖啡—配制—高等职业教育—教材—汉、英 Ⅳ.①TS273

中国国家版本馆CIP数据核字（2024）第056404号

高等职业教育餐饮类专业系列教材

咖啡（双语）

主 编 钟小立 陈丽丽 母 海
策划编辑：沈 静
责任编辑：沈 静 版式设计：沈 静
责任校对：王 倩 责任印制：张 策

＊

重庆大学出版社出版发行
出版人：陈晓阳
社址：重庆市沙坪坝区大学城西路21号
邮编：401331
电话：（023）88617190 88617185（中小学）
传真：（023）88617186 88617166
网址：http://www.cqup.com.cn
邮箱：fxk@cqup.com.cn（营销中心）
全国新华书店经销
重庆正文印务有限公司印刷

＊

开本：787mm×1092mm 1/16 印张：10.75 字数：325千
2024年5月第1版 2024年5月第1次印刷
印数：1—1 000
ISBN 978-7-5689-4331-4 定价：59.00元

前言

随着中国国际影响力不断扩大，国际化程度不断增强，社会需要大量既能懂咖啡专业知识与技能又能熟练使用英语的高素质复合型人才。本书紧密围绕行业人才需求，力求为学习者提供咖啡相关岗位所需的专业知识和技能，通过专业英语项目化系列任务丰富实践训练，提升涉外服务技能，进一步提高学习者专业核心竞争力。

为贯彻落实《国家职业教育改革实施方案》《星级饭店从业人员三年培训计划（2022—2024年）》等文件精神，根据职业教育的实际情况，结合当前经济发展趋势对餐饮、旅游、酒店类行业人才培养提出更高的要求，本书坚持"以学习者为中心，以项目任务为导向，以技能训练为基础"的原则，打破传统英语教材编写理念，以中英双语形式，构建一本集理论知识、技能训练、综合运用于一体的咖啡双语专业教材。

本书注重吸收咖啡行业的新知识、新技术、新工艺、新方法、新观念，设置了前置任务、阅读与训练、场景交际任务、知识拓展、创意任务、任务执行能力考核、目标达成考核7个模块，既能帮助学生系统掌握咖啡的基础理论和基本技能，又能帮助学生提高英语实际运用能力，实现教材的通用性、职业性、实践性、开放性和可操作性。

本书选材新颖、严谨，注重培育和践行社会主义核心价值观，特别重视专业知识传授、行业技能训练、职业素养培养三者融合，在提升学习者职业竞争力的同时，拓展国际视野，学习先进理念，丰富创新思维，促进终身发展，培养具有全球视野的新时代社会主义建设者和接班人。

本书共分中英2个部分，各有5个项目，每个项目通过若干典型任务呈现相关内容。本书具体编写分工如下：钟小立负责全书的布局、方向、统稿和审核；陈丽丽负责英文项目2任务1和项目3任务1的训练部分，以及英文项目1和项目2内容的初审；母海负责英文项目3、项目4和项目5的内容指导和初审；陈法拉负责英文项目1任务1、项目2任务2、项目3任务2和项目5任务1的训练部分；应灵弈负责英文项目1任务2和项目3任务3的训练部分；叶宁青负责英文项目4任务4的训练部分；韩梦蕾负责英文项目3任务3的训练部分；黎梓亮负责中文项目1、项目2和项目3以及视频的拍摄，吴璇负责中文项目4和项目5。

本书在编写过程中，参考了大量国内外有关专著、辞典等资料，得到了许多行业企业前辈的热心帮助与指导，特别是广州中心皇冠假日酒店、广州市炳胜饮食管理有限公司、广州市广州酒家集团职业技能培训学校有限责任公司、广州华商职业学院、广州太阳城大酒店有限公司、广州对外交流发展中心明斯克办事处、白俄罗斯国家森林公园旅游文化餐饮综合体的大力支持。重庆大学出版社编辑更是鼎力相助。我们在此表示衷心的感谢。

由于编者水平有限，书中不足之处在所难免，敬请广大读者批评指正。

编 者

2024年1月

目　录

目 录

Project Five   Café Culture

项目5　咖啡馆文化

Project One

Basic Knowledge
of Coffee

Task 1　Origin and History of Coffee

 ## Part One　Task Introduction

1. Teaching Objectives

Knowledge Objectives	1. Learn about the origin of coffee and experience the mysterious and tortuous historical journey of coffee beans. 2. Understand the basic characteristics of coffee and distinguish it from Chinese tea culture.
Skill Objectives	1. In the process of learning English, feel the way of thinking in English, and improve the logic, speculation and innovation of thinking. 2. Be able to use English to understand and express information, views and feelings more accurately, and conduct effective cross-cultural communication.
Competency Objectives	1. Cultivate the sense of collaboration and teamwork. 2. Broaden the international vision, respect the diversity of the world culture, and build a sense of community with a shared future for mankind. 3. Build confidence and set up a correct view of English learning. 4. Deepen the understanding of Chinese culture and enhance cultural self-confidence through cultural comparison.

2. Teaching Focuses

(1) Sort out the origin and history of coffee.

(2) Differences and similarities between coffee and Chinese tea.

3. Teaching Difficulty

Effective intercultural communication in English.

 ## Part Two　Task Implementation

Task Module 1　Pre-tasks

Step 1　Group Cooperation

Work in pairs. Discuss the professional terms in the box and get familiar with them. You can look them up online with your smartphones if necessary.

Ethiopia   Africa   African   Yemen   Arabia   Arab   India   Dutch

Step 2   Panel Discussion

Work in pairs to discuss the following issues and exchange views.

1. What do you know about the history of coffee?

2. What do you think is the difference between Western coffee and Chinese tea? Try to give some examples for comparison.

Task Module 2   Reading and Training

Step 1   Reading

Origin and History of Coffee

No one really knows who discovered coffee or when exactly it was discovered. But there is a wonderful legend about its origin. It is as meaningful as other creation myths. According to the legend, the coffee plant was discovered in Ethiopia by a goatherd named Kaldi around 850 AD. One evening Kaldi played his shepherd's flute and called the goats home as usual. He played it for several times but the goats didn't appear. He was confused and went to the goats. He soon found his goats very excited, bouncing back and forth. Finally, he noticed that his goats ate a kind of plant with leaves and berries which made them so excited. Kaldi tasted some of these berries. He suddenly felt all kinds of energy and excitement. He started to sing and dance around and felt like writing poetry to his girlfriend. Soon the people here recognized the energizing effect of the plant. So, the coffee plant was discovered.

It's said that at the very beginning, coffee cherries and leaves were used for invigorating people. Travelling herders in Africa mixed coffee seeds with fat and spices to create "energy bars" for long periods of time spent away from their homes. The coffee leaves and cherry skin were also boiled to create an invigorating drink.

Although there is no direct evidence, it is thought that coffee was carried to Yemen and Arabia by African slaves in the 1400s. People started to drink a refreshing potion made from coffee cherries to help them stay awake during nightly prayers. By the 1500s, Arabs had started to roast and brew the coffee seeds in a similar way to how they are prepared today.

In order to keep the monopoly and prevent others from growing coffee, coffee was strictly guarded and exported only after being cooked. So it became a precious commodity. Within a few hundred years, coffee had reached around the world, first as a beverage, then as a commodity.

 **Word and Phrase Bank**

New Words

origin /'ɒrɪdʒɪn/ *n.* 起源，起因

myth /mɪθ/ *n.* 神话

Ethiopia /ˌiːθiˈəʊpiə/ *n.* 埃塞俄比亚

goatherd /'gəʊthɜːd/ *n.* 牧羊人

flute /fluːt/ *n.* 长笛

bounce /baʊns/ *v.* （使）弹起

energizing /'enəˌdʒaɪzɪŋ/ *v.* 激励（energize的现在分词）

invigorate /ɪnˈvɪgəreɪt/ *vt.* 使精力充沛

herder /'hɜːdə/ *n.* 牧人

spice /spaɪs/ *n.* 香料，调味品

evidence /'evɪdəns/ *n.* 证据

refreshing /rɪˈfreʃɪŋ/ *adj.* 令人耳目一新的；使人凉爽的

potion /'pəʊʃn/ *n.* （药的）一服，一剂

prayer /preə(r)/ *n.* 祈祷，祷告

roast /rəʊst/ *v.* 烤，烘

monopoly /məˈnɒpəli/ *n.* 垄断，垄断权

commodity /kəˈmɒdəti/ *n.* 商品，货物

Phrases & Expressions

according to 根据

back and forth 反复地，来回地

be used for 用于，被用于做某事

in order to 为了

Step 2 After-reading Training

I Choose the best answer to complete each statement.

1. According to the legend, the coffee plant was discovered in _____.

    A. Colombia   B. Ethiopia   C. China   D. Kenya

2. Travelling herders in Africa mixed coffee seeds with _____ to create "energy bars".

    A. peanuts and raisins   B. sugar and spices

    C. fat and spices   D. fat and salt

3. People started to drink a refreshing potion made from _____ to help them stay awake.

    A. coffee cherries   B. coffee leaves   C. coffee root   D. coffee seeds

4. People drank the refreshing potion to help them stay awake during _____.

    A. sleeping   B. nightly prayers   C. task   D. studying

5. Arabs had started to _____ the coffee seeds in a similar way to how they are prepared today.

    A. grind and brew   B. roast and boil   C. roast and brew   D. dry and boil

Ⅱ   Read the statements and tick the correct boxes.

Statements	Right	Wrong	Not mentioned
Kaldi noticed that his goats ate a kind of plant with leaves and berries which made them so excited.			
At the very beginning, coffee leaves and cherry skin were ground to create an invigorating drink.			
It is thought that coffee was carried to Yemen and Arabia by Arabians in the 1400s.			
Coffee is a precious commodity today.			
The first person who commercialized coffee was a Dutch trader.			

Task Module 3   Communication Tasks

Step 1   Role-playing

A: Hi, Tom. I brought you a cup of coffee.

B: Coffee? What is it?

A: Coffee is a drink made from roasted and ground coffee beans. It's said that coffee is a global drink. It is one of the most popular drinks in the world.

B: It's darkly colored, just as black as ink. But it smells quite good.

A: Yes. Have a taste. How do you feel about it?

B: It's bitter and slightly acidic.

A: Do you like it?

B: Yes. It's quite different from our tea. But the flavor makes me happy.

A: Coffee can be prepared and presented in a variety of ways, like espresso, French press, caffè latte, or already-brewed canned coffee.

B: What kind is this, then?

A: It's espresso.

Answers to the Tasks

B: This one is hot. Is it also served cold, you know, just like other drinks?

A: Yes. Although it is usually served hot, chilled or iced coffee is common. And sugar, sugar substitutes, milk or cream are often used to lessen the bitter taste or enhance the flavor.

B: Interesting. Anything else?

A: It may be served with coffee cake or another sweet dessert, like cheesecake, tiramisu, doughnuts, etc.

B: Wow. My mouth is watering. I can't wait to have a taste. But where could I enjoy them?

A: The coffeehouse or coffee shop sells the prepared coffee beverages. How about going to Starbucks in the Central Shopping Mall now?

B: I can't agree more.

New Words

global /'gləʊbl/ *adj.*　全球的

darkly /'dɑ:kli/ *adv.*　乌黑地

acidic /ə'sɪdɪk/ *adj.*　酸的，酸性的

present /prɪ'zent/ *vt.*　（以某种方式）展现，显示

espresso /e'spresəʊ/ *n.*　（用蒸汽加压煮出的）浓缩咖啡，意式咖啡

brew /bru:/ *v.*　沏（茶），冲（咖啡）

serve /sɜ:v/ *v.*　（给某人）提供服务

chilled /tʃɪld/ *adj.*　感觉冷的，冰镇的

substitute /'sʌbstɪtju:t/ *n.*　替代品，代替物，替补队员

lessen /'lesn/ *v.*　减弱

flavor /'fleɪvə/ *n.*　风味

tiramisu /ˌtɪrəmi'su:/ *n.*　提拉米苏（一种意大利式甜点）

taste /teɪst/ *n.*　品尝，鉴赏

enhance /ɪn'hɑ:ns/ *v.*　提高，增强

coffeehouse /'kɔ:fɪˌhaʊs/ *n.*　咖啡馆，咖啡屋

Phrases & Expressions

French press　法式滤压壶，法压壶

caffè latte　拿铁咖啡

already-brewed canned coffee　罐装咖啡

sugar substitute　代糖，甜味剂

My mouth is watering.　我都流口水了。

can't wait to　等不及……

coffee shop   咖啡厅，小咖啡馆（常设在商店、旅馆等内）

Step 2   Consolidation Training

Ⅰ   Match the words on the left with their meanings on the right.

1. brew	A. smell, odor, or aroma
2. flavor	B. the sweet, usually the last course of a meal
3. enhance	C. a small restaurant where drinks and snacks are sold
4. dessert	D. to intensify or increase in quality, value, power, etc.
5. coffeehouse	E. to prepare by boiling or infusing

1. _____   2. _____   3. _____   4. _____   5. _____

Ⅱ   Decide whether the following statements are true (T) or false (F).

(   ) 1. Coffee is a brewed beverage prepared from roasted coffee leaves.

(   ) 2. Espresso can be served hot, chilled or iced.

(   ) 3. Coffee may be served with coffee cake or other desserts.

(   ) 4. Sugar, sugar substitutes, milk or cream are often used to increase the bitter taste or enhance the flavor.

(   ) 5. Only restaurants sell prepared coffee beverages.

Situation 2

Answers to the Tasks

Step 1   Role-playing

A: John, can you tell me the differences among coffee, coffee bean and coffea?

B: Coffee is the beverage. Coffea refers to the coffee tree. Coffee bean is the fruit of a coffee tree. The fruit known as coffee cherries are picked from the tree during the harvest. After processing, the coffee cherries become coffee beans.

A: I see. There are some other questions I'm very curious about. For example, when was coffee discovered?

B: Coffee was discovered at least 1, 000 years ago. But no one knows for sure.

A: Then where was it found?

B: There are two of the most prevailing species of coffee. One is Arabica. The other is Robusta. Many people believe that the origin of Arabica was in South Sudan and Ethiopia, while Robusta was born in West Africa.

A: Oh, I'm still confused.

B: I've got some information about the history of coffee. Do you want to have a look?

A: Yeah. I can't wait to see it.

 **Word and Phrase Bank**

New Words

difference /'dɪfrəns/ *n.* 差别，不同（之处）

coffea /'kɒfɪə/ *n.* 咖啡属植物

harvest /'hɑːvɪst/ *n.* 收获季节

processing /'prəʊsesɪŋ/ *n.* 加工，处理

prevailing /prɪ'veɪlɪŋ/ *adj.* 盛行的，流行的

species /'spiːʃiːz/ *n.* （动植物的）种，物种

Arabica /ə'ræbɪkə/ *n.* 阿拉比卡咖啡

Robusta /rə(ʊ)' bʌstə/ *n.* 罗布斯塔（咖啡属的一种）

confuse /kən'fjuːz/ *v.* 混淆，使困惑

Phrases & Expressions

refer to 涉及，指的是

be curious about 对……好奇

for example 例如

at least 至少

know for sure 确切地知道

South Sudan 南苏丹

West Africa 西非

Step 2 Consolidation Training

Ⅰ Complete the following sentences with the words in the box. Change the form if necessary.

> processing　discover　origin　confuse　curious

1. They decided to stick with their _____ plan.

2. Collecting data is a slow _____.

3. Children show _____ about everything.

4. We happened to _____ that we had a friend in common.

5. The instructions on the box are very _____.

Ⅱ   There are ten English words and terms numbered 1 through 10 in the left column. You should match them with their Chinese equivalents marked A through J in the right column and write down the corresponding letter in the right blank.

1. coffee bean	A. 发现
2. beverage	B. 至少
3. refer to	C. 对……好奇
4. be known as	D. 咖啡豆
5. harvest	E. 饮料
6. be curious about	F. 被称为
7. discover	G. 收获季节
8. know for sure	H. 有点儿
9. a little bit	I. 指的是
10. at least	J. 确切地知道

1. _____   2. _____   3. _____   4. _____   5. _____   6. _____
7. _____   8. _____   9. _____   10. _____

Answers to the Tasks

Task Module 4   Extensive Reading

Learn More about the Historical Transmission of Coffee

Accounts differ on the origin of the coffee plant. Although differences remain, we can trace the historical transmission of coffee with the help of both fact and legend.

1. In the middle of the 15th century the coffee tree appeared in Yemen.

2. In the 1600s coffee was exported from Yemen to other countries.

3. By the end of the 17th century, coffee had been planted in the Dutch colonies (including Indonesia).

4. In 1645 the first European coffeehouse opened in Rome.

5. In 1657 coffee was introduced to France.

6. After the 1683 Battle of Vienna coffee was planted in Austria and Poland.

7. In 1711 the first exports of Indonesian coffee from Java to the Netherlands occurred.

8. Around 1723 the first cultivation of coffee appeared in the Americas.

9. In 1727 coffee was introduced to Brazil.

10. In 1734 coffee was cultivated in Saint-Domingue (now Haiti).

11. By the end of the 18th century, coffee had become one of the world's most profitable export crops.

12. By 1920, the US consumed around half of all coffee produced worldwide.

Task Module 5   Creative Task

In order to help people more easily understand the historical transmission of coffee, you and your partners are going to draw a diagram of it. You are encouraged to collect more information from the library or through the internet if necessary.

Task 2 Species and Varieties

 ## Part One Task Introduction

1. Teaching Objectives

Knowledge Objectives	1. Learn about the classification according to the species of coffee. 2. Learn about the new varieties of coffee influenced by environmental changes.
Skill Objectives	1. In the process of learning English, feel the way of thinking in English, and improve the logic, speculation and innovation of thinking. 2. Be able to use English to understand and express information, views and feelings more accurately, and conduct effective cross-cultural communication.
Competency Objectives	1. Cultivate the sense of collaboration and teamwork. 2. Broaden the international vision, respect the diversity of the world culture, and build a sense of community with a shared future for mankind. 3. Build confidence and actively adapt to the needs of the environment. Learn to adjust the mentality in time, and establish a correct view of English learning. 4. Deepen the understanding of Chinese culture and enhance cultural self-confidence through cultural comparison.

2. Teaching Focuses

(1) Understand the species of coffee and deepen the understanding of coffee.

(2) Feel the spirit of coffee and learn to actively adapt to the environmental changes and constantly seek opportunities for survival and growth.

3. Teaching Difficulty

Effective intercultural communication in English.

 **Part Two   Task Implementation**

Task Module 1   Pre-tasks

Step 1   Group Cooperation

Work in pairs. Discuss the professional terms in the box and get familiar with them. You can look them up online with your smartphones if necessary.

> Coffea   species   Arabica   Robusta   Liberica

Step 2   Panel Discussion

Work in pairs to discuss the following issues and exchange views.

1. What species of coffee do you know?
2. What standards will farmers use to select the coffee trees they cultivate?

Task Module 2   Reading and Training

Step 1   Reading

Species and Varieties

Coffee beans come from coffee cherries while coffee cherries are from a tree known as coffea (also called coffee tree). Coffee tree ranks as one of the world's most valuable and widely traded commodity crops.

As scientists continually discover new species of coffea, the way of classifying coffea is continuously evolving. Coffea species are fond of growing wild. Nobody knows the exact quantity. At present, more than 100 species of coffea have been identified. Only the species Coffea Arabica (commonly known as "C Arabica" or "Arabica") and Coffea Robusta (known as "C. Robusta" or "Robusta") have been widely cultivated for commercial purposes, representing about 93% of global production. And Coffea Liberica (known as "C. Liberica" or "Liberica") accounts for less than 1.5%.

Why have only a few varieties been commercially grown? It's because of the various unique and inherent traits such as disease resistance and fruit yield. In order to create larger economic benefits, farmers have designed standards for the crops they would cultivate. The standards may be mainly used to select breeds for the coffee, including cupping quality, yield, resistance to diseases, resistance to pests, amount of caffeine and maturation rate.

The coffee tree is an evergreen which grows in about 70 countries. Most coffee trees grow between the Tropic of Capricorn and the Tropic of Cancer, from 25 degrees north latitude to 30 degrees south latitude, which is called the "coffee belt or coffee zone". After three to five years of cultivation, the trees can grow several meters high. As the fruits are mostly picked by hand, the trees are usually trimmed to about 1.5 m (5 ft) high to facilitate picking.

Arabica, about 60% produced worldwide, is usually bred at an altitude between 900-2,000 m (3,000-6,600 ft) above sea level. And ideally, it does best when the rainfall range is 1,500-2,500 mm (60-100 in) with an average temperature between 15-25 ℃ (60-80 °F). Robusta, nearly 40% produced worldwide, is easy to care for and has a greater crop yield. It doesn't require very cool weather and grows well in hot temperatures between 20-30 ℃ (70-85 °F). So it is usually cultivated at lower levels at an altitude between 0 and 900 m (0 and 3,000 ft) . Robustas are often planted in unstable, humid climates with frequent and heavy rainfall between 2,000-3,000 mm (80-120 in) annually.

 ## Word and Phrase Bank

New Words

commodity /kə'mɒdəti/ *n.*　商品

discover /dɪ'skʌvə(r)/ *v.*　发现

classify /'klæsɪfaɪ/ *vt.*　分类

evolve /ɪ'vɒlv/ *v.*　发展，进化

quantity /'kwɒntəti/ *n.*　数量

identify /aɪ'dentɪfaɪ/ *vt.*　识别

Arabica /ə'ræbɪkə/ *n.*　阿拉比卡（一个咖啡品种）

Robusta /ˌrə(ʊ)'bʌstə/ *n.*　罗布斯塔（一个咖啡品种）

cultivate /'kʌltɪveɪt/ *vt.*　培育

commercial /kə'mɜːʃl/ *adj.*　商业的

trait /treɪt/ *n.*　特性，特点

represent /ˌreprɪ'zent/ *vt.*　代表

Liberica /laɪ'bɪərɪəkə/ *n.*　利比里亚（一个咖啡品种）

various /'veərɪəs/ *adj.*　各种各样的

unique /ju'niːk/ *adj.*　独特的

standard /'stændəd/ *n.*　标准

altitude /'æltɪtjuːd/ *n.*　海拔高度

Phrases & Expressions

rank as　列为，排名为

coffea species　咖啡种类，咖啡属

at present　目前

global production　全球生产

account for　对……负有责任

less than　少于

inherent traits　固有特征

disease resistance　抗病性

fruit yield　果实产量

cupping quality　杯测质量

amount of caffeine　咖啡因含量

maturation rate　成熟率

pick by hand　手工采摘

rainfall range　降雨量

above sea level　海拔高度

average temperature　平均温度

Step 2　After-reading Training

I　Answer the questions.

1. Where do coffee beans come from?

_____

2. Why have only a few varieties been commercially grown?

_____

3. Why are trees usually trimmed to about 5 feet (1.5 meters) high?

_____

4. At present how many species of coffea have been identified?

_____

5. Which coffee is often planted in unstable, humid climates with frequent and heavy rainfall?

_____

II　Read the statements and tick the correct boxes.

Statements	Right	Wrong	Not mentioned
Coffea is a genus of flowering plants in the family Rubiaceae.			
The coffee tree is an evergreen which grows in about 60 countries.			
Robusta, about 40% produced worldwide, is easy to care for and has a greater crop yield.			

Continuation table

Statements	Right	Wrong	Not mentioned
As the fruits are mostly picked mechanically, the trees are usually trimmed to about 5 feet (1.5 m) high to facilitate picking.			
The traits which may be mainly used to select breeds for coffee include cup quality, yield, resistance to diseases, resistance to pests, amount of caffeine and maturation rate.			

Task Module 3   Communication Tasks

Answers to the Tasks

Situation 1

Step 1   Role-playing

A: Kayla, Willa told me that a coffee fruit contains two coffee seeds with their flat sides together.

B: Yes. The coffee fruit is also called cherry while the coffee seeds are called coffee beans.

A: OK. But I'm curious if there is any coffee cherry that only contains one single seed, instead of two.

B: In fact, there is one kind of coffee cherry known as peaberry. The peaberry only has a single seed which develops with nothing to flatten it. So it is usually oval or pea-shaped. Only around 5% of all coffee beans harvested are of this form.

A: Is peaberry more delicious?

B: I don't know. It's said that these beans have more flavor than other normal beans as they consist mostly of endosperm.

A: Interesting. I really want to have a try. By the way, do you know the DNA genetic authentication of plant material?

B: I've just read some information about these technologies. It's here. It says that in 2020 this technique was proven effective for coffee.

A: What's it for?

B: It may help farmers to improve the coffee production. It also enhances transparency and traceability of coffee beans.

A: Excellent. In this way, it will be safer to buy foreign raw coffee beans online in the future.

Word and Phrase Bank

New Words

contain /kən'teɪn/ *vt.*　容纳，包含

flat /flæt/ *adj.*　平的，平滑的

peaberry /'piːberi/ *n.*　珠粒（一荚单粒的圆粒咖啡豆）

flatten /'flætn/ *v.*　压平，(使)变平

oval /'əʊvl/ *adj.*　椭圆形的

flavor /'fleɪvə(r)/ *n.*　味道，风味

normal /'nɔːml/ *adj.*　正常的

endosperm /'endəʊspɜːm/ *n.*　[植]胚乳

genetic /dʒə'netɪk/ *adj.*　遗传的，基因的

authentication /ɔː,θentɪ'keɪʃn/ *n.*　认证，鉴定

technology /tek'nɒlədʒɪ/ *n.*　技术，工艺

effective /ɪ'fektɪv/ *adj.*　有效的

transparency /træns'pærənsɪ/ *n.*　透明度

traceability /treɪsə'bɪlɪti/ *n.*　可追踪性

online /,ɒn'laɪn/ *adv.*　在线地

Phrases & Expressions

instead of　作为……的替换，而不是

pea-shaped　豌豆形的

consist of　包括，由……组成

have a try　试试

by the way　顺便说一句

DNA genetic authentication　DNA基因鉴定

raw coffee bean　咖啡生豆

in the future　未来

Step 2　Consolidation Training

Ⅰ　Match the words on the left with their meanings on the right.

1. genetic	A. a practical method or art applied to some particular task
2. technique	B. to become or make something become flat or flatter
3. flatten	C. the quality of being clear and transparent

Continuation table

| 4. transparency | D. a substance that things can be made from |
| 5. material | E. tending to occur among members of a family usually by heredity |

1. _____   2. _____   3. _____   4. _____   5. _____

Ⅱ Decide whether the following statements are true (T) or false (F).

(   ) 1. A coffee fruit contains three coffee seeds with their flat sides together.

(   ) 2. It will be safer to buy foreign raw coffee beans online in the future.

(   ) 3. It's said that the peaberries have more flavor than other normal beans as they consist mostly of endosperm.

(   ) 4. The DNA genetic authentication in 2020 was proven ineffective for coffee.

(   ) 5. The coffee fruit is also called cherry while the coffee seeds are called coffee beans.

Situation 2

Answers to the Tasks

Step 1   Role-playing

A: Kayla, which place mostly produces Robusta in recent years?

B: Vietnam. And Vietnam has become the world's largest exporter of Robusta coffee. Vietnam accounts for over 40% of the total production, while Brazil 25%, Indonesia 13%, India 5% and Uganda 5%. The coffee production in Vietnam surpasses that in Brazil.

A: Brazil is the biggest coffee producer in the world, isn't it?

B: Yes, it is. It does produce one-third of the world's coffee. But about 70% of that is Arabica.

A: Do you know what the differences between Arabica and Robusta are?

B: Arabica has a smoother taste with more acidity and a richer flavor while Robusta is less acidic, more bitter and often has a distinctive earthy taste.

A: How about their caffeine content?

B: We can find out the answer from this picture. Arabica beans contain 0.8%-1.4% caffeine and 6%-9% sugar while Robusta has 1.7%-4% caffeine and 3%-7% sugar.

A: More caffeine and less sugar. I now understand why Robusta is bitter. In addition to introducing the species of coffee, is there any information about the most famous coffee beans in the world?

B: Yes. For example, many people think that Blue Mountain Coffee is the "King of Coffee" and Hawaiian Kona Coffee is the "Queen of Coffee". These coffees are rare and expensive. The Brazilian Santos is the representative of high-quality Brazilian coffee. And there are some

information about other coffee such as Colombian Premium Coffee, Indonesian Kopi Luwak and Mandeling Coffee.

 Word and Phrase Bank

New Words

mostly /ˈməʊstli/ *adv.* 主要地

recent /ˈriːsnt/ *adj.* 最近的

Vietnam /ˌvjetˈnɑːm/ *n.* 越南

exporter /ekˈspɔːtə(r)/ *n.* 输出国

percentage /pəˈsentɪdʒ/ *n.* 百分率，百分比

total /ˈtəʊtl/ *adj.* 总计的

Brazil /brəˈzɪl/ *n.* 巴西

Indonesia /ˌɪndəˈniːʒə/ *n.* 印度尼西亚

India /ˈɪndɪə/ *n.* 印度

Uganda /juːˈgændə/ *n.* 乌干达

surpass /səˈpɑːs/ *v.* 超越，超过

producer /prəˈdjuːsə(r)/ *n.* 产地，生产商，制造商

one-third /wʌnθɜːd/ *n.* 三分之一

smooth /smuːð/ *adj.* 顺滑的，醇和的

distinctive /dɪˈstɪŋktɪv/ *adj.* 独特的

earthy /ˈɜːθɪ/ *adj.* 土的，泥土的

content /ˈkɒntent/ *n.* 含量

representative /ˌreprɪˈzentətɪv/ *n.* 代表

Phrases & Expressions

total production　总产量

a distinctive earthy taste　独特的泥土味

from this picture　从这张图中

find out　发现

in addition to　除……之外

Blue Mountain Coffee　蓝山咖啡

Brazilian Santos　巴西桑托斯咖啡

Colombian Premium Coffee　哥伦比亚优质咖啡

Indonesian Kopi Luwak　印度尼西亚猫屎咖啡

Mandeling Coffee　曼特宁咖啡

Step 2 Consolidation Training

Ⅰ Complete the following sentences with the words in the box. Change the form if necessary.

> account   find   like   contain   regard

1. Vietnam _____ for over 40% of the total production, while Brazil 25%, Indonesia 13%, India 5% and Uganda 5%.
2. You can _____ out the percentages of Robusta coffee produced in different places.
3. There is some information about other coffee _____ Colombian Premium Coffee, Indonesian Kopi Luwak and Mandeling Coffee.
4. Arabica beans _____ 0.8%-1.4% caffeine and 6%-9% sugar while Robusta has 1.7%-4% caffeine and 3%-7% sugar.
5. Blue Mountain Coffee from Jamaica is _____ as the best coffee, rare and expensive.

Ⅱ There are ten English words and terms numbered 1 through 10 in the left column. You should match them with their Chinese equivalents marked A through J in the right column and write down the corresponding letter in the right blank.

1. representative	A. 从这张图中
2. in addition to	B. 质量
3. a distinctive earthy taste	C. 内容
4. from this picture	D. 代表
5. expensive	E. 昂贵的
6. understand	F. 品尝
7. content	G. 除……之外
8. quality	H. 独特的泥土味
9. species	I. 理解，明白
10. taste	J. 种类

1. _____   2. _____   3. _____   4. _____   5. _____   6. _____
7. _____   8. _____   9. _____   10. _____

Answers to the Tasks

Task Module 4 Extensive Reading

Learn More about the Classification of Coffee

There are many ways to classify coffee. One of the most commonly used ways is to classify coffee according to the original species of Coffea. There are over 120 species of Coffea and the two most popular are Arabica and Robusta. Arabica coffee has its origin in Ethiopia and was first cultivated in Yemen. Climate change such as rising temperatures, longer droughts, and excessive rainfall will threaten the sustainability of Arabica coffee production and makes the Coffea more difficult to cultivate. The best-known Arabica coffee beans are from Jamaican Blue Mountain, Colombian Supremo, Tarrazú, Costa Rica, Guatemalan Antigua, and Ethiopian Sidamo.

Robusta coffee originates from central and western sub-Saharan Africa. Compared to Arabica coffee, Robusta coffee is easier to care for and is less susceptible to pests and diseases. This leads to a higher yield than that of Arabica. Robusta is mostly grown in Vietnam but also produced in India, Africa and Brazil.

Other species of coffee exist and contain unique varieties. For instance, Kapeng Barako, a Liberica variety grown in the Philippines, has a strong flavor and fragrance reminiscent of aniseed. Coffea Charrieriana is a caffeine-free coffee found in Cameroon. Coffea Stenophylla can grow at higher temperatures than Arabica and has a better flavor profile than Robusta.

We can also classify coffee by the taste, including acidity, bitterness, sweetness, saltiness and sourness. Another method of classification is by region, such as coffee from Ethiopia, Jamaica, Yemen, Costa Rica, Guatemala, Africa, Indonesia, Caribbean, Brazil, etc.

Task Module 5 Creative Project

In order to better feel the difference between Arabica and Robusta, please try to get the two kinds of coffee beans with your partner. Then compare them and try to share your ideas with other students in the class. You are encouraged to collect more reference information from the library or through the internet if necessary.

项目 1

咖啡基础知识

任务1　咖啡的起源和历史

第1部分　任务介绍

咖啡的起源和历史

1. 教学目标

知识目标	1. 了解咖啡的起源，感受咖啡豆经历的神秘且曲折的历史旅途。 2. 了解咖啡的基本特点，以及与中国茶文化的区别。
技能目标	1. 在学习英语的过程中，感受英语思维方式，提升思维的逻辑性、思辨性和创新性。 2. 能够运用英语比较准确地理解和表达信息、观点、情感，能够有效地进行跨文化交际与沟通。
素质目标	1. 培养与人协作的精神和团队合作意识。 2. 拓宽国际视野，尊重多元文化，树立人类命运共同体意识。 3. 树立信心，树立正确的英语学习观。 4. 通过文化对比，加深对中华传统文化的理解，增强文化自信。

2. 教学重点

（1）咖啡起源与历史的梳理。

（2）咖啡与中国茶的异同。

3. 教学难点

能有效地用英语进行跨文化交际与沟通。

第2部分　任务实施

任务模块1　前置任务

步骤1　小组合作

两人一组，讨论下列方框中的专业用语，并加以熟悉。如果需要，可以使用智能手机在线查找。

> Ethiopia　Africa　African　Yemen　Arabia　Arab　India　Dutch

步骤2 小组讨论

两人一组，讨论以下问题并交流观点。

1. 你对咖啡的历史了解多少？

2. 你认为咖啡和中国的茶有什么不同？试着举例比较两者的差异。

任务模块2 阅读与训练

咖啡的起源和历史

没有人真正知道咖啡是谁发现的，也没有人知道咖啡是什么时候被发现的。但是，关于咖啡的起源有一个奇妙的传说。咖啡与其他创造神话一样有意义。据说，大约在公元850年，一个名叫卡尔迪的牧羊人在埃塞俄比亚发现了一种咖啡树。一天晚上，卡尔迪吹着他的牧羊笛子，像往常一样把山羊叫回家。他吹了好几次，但是山羊都没有出现。他觉得很奇怪，就去找山羊。很快，他发现他的山羊非常兴奋，不停蹦蹦跳跳。这时，卡尔迪注意到他的山羊吃了一种树叶和浆果，让它们非常兴奋。卡尔迪尝了一些浆果，他突然感到精力充沛，异常兴奋。他开始又唱又跳，还想给女朋友写诗。很快，这里的人们认识到这种植物有刺激和提神的功效。咖啡树就此被发现了。

据说一开始，咖啡的果实和叶子是用来提神的。在非洲旅行的牧民将咖啡的种子、脂肪和香料混合在一起，为长期远离家园的人们创造"能量棒"。咖啡的叶子和咖啡的果皮也被煮沸，制成一种有活力的饮料。

虽然无从考证，但是人们还是认为咖啡是在15世纪由非洲传到也门和阿拉伯的。人们开始喝由咖啡果皮制成的提神饮品，以帮助他们在夜间祈祷时保持清醒。到了16世纪，阿拉伯人开始用与今天相似的方式烘焙和冲泡咖啡。

为了保持对咖啡的垄断，防止其他人种植咖啡，咖啡被严格保护起来，只有煮熟后才能出口。因此，咖啡成了珍贵的商品。在之后的几百年里，咖啡首先作为饮料，然后作为商品传播到世界各地。

任务模块3 场景交际任务

场景1

A：嗨，汤姆。我给你带了一杯咖啡。

B：咖啡？是什么？

A：咖啡是一种用经过烘焙和研磨的咖啡豆制作出来的饮品。据说咖啡是一种全球性的饮品。它是世界上最受欢迎的饮品之一。

B：这个颜色这么黑，就像墨水一样。不过闻起来好香。

A：对，尝一尝。你觉得怎么样？

B：这是苦的，还有点酸。

A：你喜欢吗？

B：喜欢。这和我们的茶很不一样。但是这个味道让我很开心。

A：咖啡可以以多种方式制作和冲泡，如意式咖啡、法压壶咖啡、拿铁或即饮罐装咖啡。

B：那这杯是什么咖啡？

A：这杯是意式咖啡。

B：这杯是热的。它可以像其他饮品一样做成冷饮吗？

A：可以的。虽然通常是提供热的咖啡，但是冷冻的或冰咖啡也很常见。通常还会用糖、代糖、牛奶或奶油来减轻苦味或增强风味。

B：太有趣了。还有别的吗？

A：咖啡可以搭配咖啡蛋糕或其他甜点，如奶酪蛋糕、提拉米苏、甜甜圈等。

B：哇，我都流口水了，好想尝一尝。但是哪里有卖呢？

A：咖啡馆或咖啡店有卖咖啡饮品。我们现在去中央购物中心的星巴克怎么样？

B：太好了。

场景2

A：约翰，你能告诉我咖啡、咖啡豆和咖啡属植物的区别吗？

B：咖啡是饮品，咖啡属植物指的是咖啡树。咖啡豆是咖啡树的果实。这种果实被称为咖啡果（又称咖啡樱桃），是在收获期间从树上摘下来的。经过加工，咖啡果会变成咖啡豆。

A：我明白了。还有一些问题我感到很好奇。例如，咖啡是什么时候发现的？

B：咖啡大约在1000年前被发现，但没有人知道确切时间。

A：咖啡是在哪里被发现的？

B：有两个最流行的咖啡品种：一个是阿拉比卡；另一个是罗布斯塔。很多人认为，阿拉比卡起源于南苏丹和埃塞俄比亚，而罗布斯塔起源于西非。

A：哦，我还是很困惑。

B：我有一些关于咖啡历史的信息。你想看看吗？

A：想，我都等不及要看了。

任务模块4　知识拓展

了解更多关于咖啡历史传承的信息

关于咖啡的起源有不同的说法。尽管说法不一，但是，我们可以借助事实和传说来追溯咖啡的历史传承。

1. 15世纪中期，咖啡树出现在也门。

2. 17世纪，咖啡从也门出口到其他国家。

3. 17世纪末，咖啡已经开始在荷兰殖民地（包括印度尼西亚）种植。

4. 1645年，第一家欧洲咖啡馆在罗马开业。

5. 1657年，咖啡被引入法国。

6. 1683年，维也纳战役后，咖啡在奥地利和波兰种植。

7. 1711年，印度尼西亚咖啡首次从爪哇出口到荷兰。

8. 1723年前后，美洲出现了第一批咖啡的种植。

9. 1727年，咖啡被引入巴西。

10. 1734年，咖啡在圣多明各（现在的海地）种植。

11. 18世纪末，咖啡已成为世界上最赚钱的出口作物之一。

12. 1920年，美国消费了全世界产量一半左右的咖啡。

任务模块5　创意任务

为了帮助人们更容易理解咖啡的历史传承，你和你的合作伙伴将一起绘制一个图表。我们鼓励你在必要时从图书馆或通过互联网获取更多信息。

任务模块6　任务执行能力考核

序号	考核细分项目	标准分数（分）	得分（分）
1	前置任务	10	
2	咖啡历史溯源任务	20	
3	场景交际任务1	20	
4	场景交际任务2	20	
5	知识拓展	10	
6	创意任务	20	

任务模块7　目标达成考核

考核项目＼评分	标准分数（分）	个人自评	小组互评	教师评分
理论知识	30			
技能训练	40			
职业素养	30			
总分（分）	100			
综合总分				
说明	综合总分=个人自评（占总分的20%）+小组互评（占总分的20%）+教师评分（占总分的60%）			

任务2 咖啡种类

 ## 第1部分 任务介绍

咖啡种类

1. 教学目标

知识目标	1. 了解咖啡的种类和分类方法。 2. 了解咖啡为适应环境变化而产生的变种情况。
技能目标	1. 在学习英语过程中，感受英语思维方式，提升思维的逻辑性、思辨性、创新性。 2. 能够运用英语，比较准确地理解和表达信息、观点、情感，进行有效的跨文化交际与沟通。
素质目标	1. 培养与人协作的精神和团队合作的意识。 2. 拓宽国际视野，尊重世界多元文化，树立人类命运共同体意识。 3. 树立信心，积极适应环境需求，学会及时调整心态，树立正确的英语学习观。 4. 通过文化对比，加深对中华文化的理解，增强文化自信。

2. 教学重点

（1）了解咖啡的种类，加深对咖啡的认识。

（2）感受咖啡积极适应环境、不断寻求生存与成长机会的精神。

3. 教学难点

能用英语进行跨文化交际与沟通。

 ## 第2部分 任务实施

任务模块1 前置任务

步骤1 小组合作

两人一组，讨论下列方框中的专业用语并加以熟悉。如果需要，可以用智能手机在线查找。

> Coffea　species　Arabica　Robusta　Liberica

步骤2　小组讨论

两人一组，讨论以下问题并交流观点。

1. 你知道的咖啡种类有哪些？
2. 农民用什么标准来选择他们种植的咖啡树？

任务模块2　阅读与训练

咖啡种类

咖啡豆来自咖啡果，而咖啡果来自一种名为咖啡属植物（Coffea）的树（也称咖啡树）。咖啡树是世界上最有价值和交易最广泛的商品作物之一。

随着科学家不断发现新的咖啡种类，咖啡的分类方法也在不断发展。咖啡树喜欢在野外生长。没有人知道确切的数量。目前，已鉴定出的咖啡树有100多个种类。只有阿拉比卡（通常称为"阿拉比卡咖啡"或"阿拉比卡"）和罗布斯塔（通常称为"罗布斯塔咖啡"或"罗布斯塔"）被广泛用于商业目的，约占全球产量的93%。咖啡属利比里亚（通常称为"利比里亚咖啡"或"利比里亚"）所占比例不到1.5%。

为什么只有少数种类的咖啡被商业化种植？这是因为各个种类的咖啡有着独特的、固有的特性，如抗病性和果实产量。为了创造更高的经济效益，农民为他们种植的作物制定了标准。这些标准主要用于选择咖啡种类，包括杯测质量、产量、抗病性、抗虫性、咖啡因含量和成熟率。

咖啡树是一种常绿植物，生长在大约70个国家。咖啡树大多生长在南北回归线之间，从北纬25°到南纬30°，被称为"咖啡生长带"。经过3～5年的栽培，这些咖啡树可以长到数米高。由于大部分果实都是手工采摘的，因此通常会将树木修剪至约1.5米（5英尺）高，以便于采摘。

阿拉比卡在全球的产量约占60%，通常在海拔900～2000米（3000～6600英尺）的地方繁殖。降雨量在1500～2500毫米（60～100英寸），平均温度在15～25 ℃（60～80 °F）为理想状态，种植效果最好。罗布斯塔（Robusta）在全球的产量接近40%，这个品种容易护理，产量更高。它不需要非常凉爽的天气，在20～30 ℃（70～85 °F）的高温下生长良好。因此，它通常在海拔0～900米（0～3000英尺）的较低水平上种植。罗布斯塔通常种植在年降雨量为2000～3000毫米（80～120英寸）的不稳定的、潮湿的气候中。

任务模块3　场景交际任务

场景1

A：凯拉，薇拉跟我说，一个咖啡果里有两个扁平的果核。
B：对。咖啡果也叫樱桃果，咖啡果核叫咖啡豆。

A：好。我很好奇，是否有咖啡果里只含有一颗果核，而不是两颗。

B：事实上，有一种咖啡果叫小圆豆。小圆豆只有一颗种子，没有任何东西可以使它变平。所以，它通常是椭圆形或豌豆形。所有收获的咖啡豆中，只有大约5%是这种形状的。

A：小圆豆更美味吗？

B：我不知道。据说这些豆子比其他普通豆子有更多的味道，因为它们主要由胚乳组成。

A：真有趣。我好想尝一尝。顺便问一下，你知道植物原料的DNA基因鉴定吗？

B：我刚好读了一些关于这些技术的信息。就在这里。这里说，在2020年这项技术被证明对咖啡是有效的。

A：为什么？

B：它可以帮助农民提高咖啡产量。它还能使咖啡豆的透明度和可追踪性增强。

A：好极了。这样的话，未来在网上购买外国咖啡生豆将会更加安全。

A：凯拉，近年来，哪个地方主要生产罗布斯塔？

B：越南。越南已成为世界上最大的罗布斯塔咖啡出口国。越南占总产量的40%以上，巴西占25%，印度尼西亚占13%，印度占5%，乌干达占5%。越南的罗布斯塔咖啡产量超过了巴西。

A：巴西是世界上最大的咖啡生产国，不是吗？

B：是的。它确实生产了世界1/3的咖啡，但其中约70%是阿拉比卡咖啡。

A：你知道阿拉比卡和罗布斯塔有什么不同吗？

B：阿拉比卡口感更顺滑，酸度更高，味道更丰富；而罗布斯塔酸度更低、更苦，通常带有独特的泥土味。

A：它们的咖啡因含量如何？

B：我们可以从这张图中发现答案。阿拉比卡咖啡豆含有0.8%~1.4%的咖啡因和6%~9%的糖，而罗布斯塔咖啡豆含有1.7%~4%的咖啡碱和3%~7%的糖。

A：更多的咖啡因和更少的糖。我现在明白为什么罗布斯塔是苦的了。除了介绍咖啡的种类，还有关于世界上最著名的咖啡豆的信息吗？

B：有的。例如，很多人认为蓝山咖啡是"咖啡皇帝"，夏威夷科纳咖啡是"咖啡皇后"，这些咖啡稀有且昂贵。巴西桑托斯是高品质巴西咖啡的代表。还有一些关于其他咖啡的信息，如哥伦比亚优质咖啡、印度尼西亚猫屎咖啡和曼特宁咖啡。

任务模块4　知识拓展

了解更多关于咖啡分类的信息

咖啡的分类方法很多。最常用的方法之一是根据咖啡树的原始种类对咖啡进行分类。有120多种咖啡树，其中，最受欢迎的两种是阿拉比卡咖啡和罗布斯塔咖啡。阿拉比卡咖啡起源于埃塞俄比亚，最初在也门种植。气候变化，如气温上升、干旱时间延长和降雨量过多，会威胁阿拉比卡咖啡生产的可持续性，使这种咖啡更难种植。最著名的阿拉比卡咖啡豆来自牙买加的蓝山、哥伦比亚、哥斯达黎加、危地马拉安提瓜和埃塞俄比亚西达摩。

罗布斯塔咖啡起源于撒哈拉以南的非洲中部和西部。与阿拉比卡咖啡相比，它更容易养护，更不容易受虫害和疾病的影响。因此，罗布斯塔咖啡比阿拉比卡咖啡有更高的产量。罗布斯塔咖啡主要种植在越南，但在印度、非洲和巴西也有生产。

其他种类的咖啡也是有的，它们有独特的品种。例如，生长在菲律宾的利比里亚品种八打雁咖啡具有强烈的味道和香味，会令人联想到茴香。卡里尔咖啡树是一种在喀麦隆发现的无咖啡因的品种。与阿拉比卡咖啡相比，狭叶咖啡可以在更高的温度下生长，与罗布斯塔咖啡相比，具有更好的风味。

我们还可以根据咖啡的味道来分类，包括酸度、苦味、甜度、咸度和酸味。另一种分类方法是根据地区分类，如埃塞俄比亚、牙买加、也门、哥斯达黎加、危地马拉、非洲、印度尼西亚、加勒比、巴西等。

任务模块5　创意任务

为了更好地感受阿拉比卡咖啡和罗布斯塔咖啡的区别，请尝试与你的组员一起购买这两种咖啡豆。然后进行比较，并尝试与课堂上的其他学生分享你们的想法。如有必要，我们鼓励你们从图书馆或通过互联网收集更多参考信息。

任务模块6　任务执行能力考核

序号	考核细分项目	标准分数（分）	得分（分）
1	专业用语	10	
2	任务资料的基础训练	20	
3	场景任务1专项训练	20	
4	场景任务2专项训练	20	
5	知识拓展	10	
6	创意任务	20	

任务模块7　目标达成考核

评分＼考核项目	标准分数（分）	个人自评	小组互评	教师评分
理论知识	30			
技能训练	40			
职业素养	30			
总分（分）	100			
综合总分				
说明	综合总分=个人自评（占总分的20%）+小组互评（占总分的20%）+教师评分（占总分的60%）			

Processing

Task 1   Processing of Coffee Fruit

 Part One   Task Introduction

1. Teaching Objectives

Knowledge Objectives	1. Learn about coffee fruit picking and common processing methods. 2. Understand the packaging characteristics of green coffee beans. 3. Understand the market demand for green coffee beans.
Skill Objectives	1. Combining theory with practice, learn to draw a flow chart of coffee fruit processing. 2. In the process of learning English, feel the way of thinking in English, and improve the logic, speculation and innovation of thinking. 3. Be able to use English to understand and express information, views and feelings more accurately, and conduct effective cross-cultural communication.
Competency Objectives	1. Cultivate the sense of collaboration and teamwork. 2. Broaden the international vision, respect the diversity of the world culture, and build a sense of community with a shared future for mankind. 3. Build confidence and actively adapt to the needs of the environment. 4. Learn to adjust the mentality in time, and establish a correct view of English learning. Train logical thinking and information processing ability, and cultivate creative consciousness.

2. Teaching Focuses

(1) Processing methods of coffee fruits.

(2) The function and influence of green coffee bean packaging.

3. Teaching Difficulty

Effective intercultural communication in English.

Part Two　Task Implementation

Task Module 1　Pre-tasks

Step 1　Group Cooperation

Work in pairs. Discuss the professional terms in the box and get familiar with them. You can look them up online with your smartphones if necessary.

> cherry　berry　harvest　processing　labor-intensive　density　mass consumer

Step 2　Panel Discussion

Work in pairs to discuss the following issues and exchange views.

1. Do you know how the coffee fruit is picked?
2. How many steps do you know about the coffee fruit processing and what are they?

Task Module 2　Reading and Training

Step 1　Reading

Processing of Coffee Fruit

In order to become coffee beans, the fruit of a coffee tree (commonly known as coffee cherries or berries) needs to be processed. But first, all the cherries should be collected from the tree.

Generally speaking, there are two ways to collect cherries. The first way is called strip-picking. All coffee fruit is removed from the tree with mechanical strippers, regardless of maturation state. This method is often used in the medium and large farms located in Brazil and Australia with dry climate and immense coffee fields. The second way is called hand-picking. In most countries, only cherries at the peak of ripeness are picked individually by hand. As this kind of harvest is labor-intensive and more costly, it is mainly used to harvest higher quality Arabica beans.

After picking, all the harvested cherries should be processed within a few hours to preserve their quality. Before the dry mill stage, there are two main methods to process the cherries, including natural/dry-process and washed/wet processed. With these methods, the cherries will be turned into the shrunk brown dry coffee beans or the parchment-covered coffee beans. These beans then will be transferred into a dry mill which will help to remove their dried skin and parchment. The sorting could be done by the dry mill or by hand.

There are several ways to sort green coffee: by weight, size and/or color. The low- or average-quality green coffee beans will be poured into containers for producing cheaper mass consumer coffee. The beans of the best quality will be shipped to the specialty market. The traditional jute bags or woven poly bags are extremely porous that will expose the green coffee to its surroundings. This poor storage may develop the beans a burlap-like taste and its positive qualities may fade away. In order to preserve the quality of green coffee, in recent years the specialty coffee market has begun to choose some storage performance enhancing methods, like storing the green coffee in protective bags lined with plastic or in vacuum packaging. The vacuum packaging could further reduce the ability of green coffee to be affected by oxygen under atmospheric humidity. But it is more expensive.

It's a long journey for coffee from a farm to a roasting machine. Simply speaking, it contains planting, picking, screening, dry-milling processing and transportation. After the above steps the coffee beans buyers will roast, grind and brew the beans before they serve the drinks to others.

Word and Phrase Bank

New Words

mechanical /mə'kænɪkl/ *adj.*　机械（方面）的

stripper /'strɪpə(r)/ *n.*　剥离器

regardless /rɪ'gɑːdləs/ adv.　不顾，不加理会

immense /ɪ'mens/ *adj.*　极大的，巨大的

ripeness /'raɪpnəs/ *n.*　成熟

individually /ˌɪndɪ'vɪdʒuəli/ adv.　分别地，单独地

harvest /'hɑːvɪst/ *n.*　收获；*v.*　收获，收割

Arabica /ə'ræbɪkə/ *n.*　阿拉比卡咖啡豆

preserve /prɪ'zɜːv/ *v.*　保持，维持

shrink /ʃrɪŋk/ *v.*　收缩

specialty /'speʃəlti/ *n.*　专业，专长

woven /'wəʊvn/ *adj.*　织物的

porous /'pɔːrəs/ *adj.*　能渗透的，有气孔的

atmospheric /ˌætməs'ferɪk/ *adj.*　大气的

Phrases & Expressions

strip-picking　速剥采收法

maturation state　成熟的状态

hand-picking　手工采摘

labor-intensive　劳动密集型的

dry mill　干磨机

natural/dry-process　日晒法

washed/wet processed　水洗法

pour into　涌入

mass consumer　大众消费者

fade away　逐渐消失

jute bag　黄麻袋

vacuum packaging　真空包装

atmospheric humidity　大气湿度

simply speaking　简单地说

Step 2　After-reading Training

Ⅰ　Choose the best answer to complete each statement.

1. In order to become coffee beans, the fruit of a coffee plant needs to be _____.

    A. boiled　　　　　B. processed　　　　　C. dried　　　　　D. milled

2. Strip-picking is often used in the medium and large farms located in _____.

    A. Brazil and America　　　　　　　B. Chile and Mexico

    C. New Zealand and Australia　　　　D. Brazil and Australia

3. Hand-picking is mainly used to harvest higher quality _____.

    A. Robusta beans　　B. Arabica beans　　C. Colombia beans　　D. Liberica beans

4. The beans will be sorted into low- to high-quality by _____, size and/or color.

    A. density　　　　　B. odor　　　　　C. weight　　　　　D. milled

5. The traditional jute bags or _____ are extremely porous that will expose the green coffee to its surroundings.

    A. wooden barrels　　B. vacuum bags　　C. woven poly bags　　D. plastic bags

Ⅱ　Read the statements and tick the correct boxes.

Statements	Right	Wrong	Not mentioned
The first way to collect coffee fruit is called strip-picking, which coffee fruit is removed from the tree with mechanical strippers.			
Strip-picking is often used in the small farms located in Brazil and Australia with dry climates and immense coffee fields.			
In most countries, all cherries are picked by hand.			
After picking, all the harvested cherries should be processed within a few hours to preserve their quality.			

Continuation table

Statements	Right	Wrong	Not mentioned
The average-quality green coffee beans will be poured into containers for the specialty market.			

Task Module 3　Communication Tasks

Situation 1

Answers to the Tasks

Step 1　Role-playing

A: Christina, I've got some confusion about the dry-milling process of coffee cherries. Could you help me?

B: I'll have a try. What's your question?

A: Who will dry and mill the coffee cherries? The farmer or someone else?

B: As I know, some farmers owning their own mills will process the cherries themselves so that they can keep quality control of the green coffee until export. While other producers without their own mills may sell the cherries to centralize "stations" where the cherries will be dried and milled.

A: I see. After the milling, what will be done to the coffee beans?

B: They will be divided into different categories which will indicate the quality of the beans. People often say that in the coffee world any coffee bean, from the cheapest ones to the top 1 percent, has its buyer.

A: Why?

B: The green beans of low quality will be sold at the lowest price for producing cheaper mass consumer coffee. They are characterized by a displeasingly bitter or astringent flavor and a sharp odor.

A: How about the best ones?

B: The beans of the best quality will be more expensive as they are more fragrant, smooth, and mellow with their higher aromatic oil but lower organic acid content.

A: Just as what we often say, the price is determined by quality.

Word and Phrase Bank

New Words

confusion /kən'fjuːʒn/ *n.*　困惑，不明确

export /'ekspɔːt/ *v.*　出口；传播，输出

centralize /'sentrəlaɪz/ *v.*　集权控制，使集中

category /'kætəgəri/ *n.*　种类，范畴

indicate /'ɪndɪkeɪt/ *v.* 表明，标示

characterize /'kærəktəraɪz/ *v.* 以……为特征

astringent /ə'strɪndʒənt/ *adj.* 涩的

mellow /'meləʊ/ *adj.* （水果）成熟香甜的

aromatic /ˌærə'mætɪk/ *adj.* 芳香的，芬芳的

Phrases & Expressions

divide into　把……分成

just as　正如

Step 2　Consolidation Training

Ⅰ　Match the words on the left with their meanings on the right.

1. mill	A. give evidence of
2. centralized	B. a degree or grade of excellence or worth
3. indicate	C. to settle or decide by choice of alternatives or possibilities
4. category	D. the act of grinding to a powder or dust
5. quality	E. a person who uses goods or services
6. consumer	F. concentrated on or clustered around a central point or purpose
7. astringent	G. pleasant-smelling
8. displeasingly	H. a collection of things sharing a common attribute
9. fragrant	I. sour or bitter in taste
10. determine	J. in a displeasing manner

1. _____　2. _____　3. _____　4. _____　5. _____　6. _____　7. _____

8. _____　9. _____　10. _____

Ⅱ　Decide whether the following statements are true (T) or false (F).

(　　) 1. Some farmers who own their own mills will process the cherries themselves.

(　　) 2. Coffee beans are divided into different categories which will indicate the size of the beans.

(　　) 3. The green beans of high quality will be sold at a high price for producing mass consumer coffee.

(　　) 4. The green beans of low quality are characterized by a displeasingly bitter or astringent flavor and a sharp odor.

(　　) 5. The beans of the best quality will be more expensive as they are less fragrant.

Answers to the Tasks

Situation 2

Step 1   Role-playing

A: Teresa, why do you want to change the packaging of coffee beans with item No. 225?

B: No. 225 is one of our most expensive products as it has the top quality. If we use the jute bags, this poor package will just create a burlap-like taste and decline its positive quality. In that case, how can we sell it at a higher price?

A: But if you change the jute bag to vacuum packaging, the cost will be increased immediately.

B: Don't worry about it. We've surveyed our several focus consuming groups. The result shows that they preferred better package for beans with superior quality. The increase in price is acceptable.

A: I'm still worried about the price. How much percentage of price increase would be acceptable?

B: How about having a trial run? Meanwhile we could try to create our new brand with our new package.

A: Good idea. So what do you want to do?

B: I'm thinking about the packing method of Blue Mountain Coffee.

A: What do you mean?

B: Unlike most green coffees being packaged and transported in cloth bags of 60 kg / bag, Blue Mountain Coffee are stored, shipped and sold in wooden barrels of 70 kg / barrel. And as we know, Jamaica is the last country still transports coffee in traditional wooden barrels. When we buy Blue Mountain Coffee beans from Jamaica, the package is one of their selling points.

A: Indeed. I know what you're thinking now.

 Word and Phrase Bank

New Words

immediately /ɪ'miːdiətli/ *adv.*   立即，马上
survey /'sɜːveɪ/ *vt.*   考察，调查
superior /suː'pɪəriə(r)/ *adj.*   高质量的，优质的
acceptable /ək'septəb(ə)l/ *adj.*   可接受的，（大多数人）认同的

Phrases & Expressions

trial run   试验
Blue Mountain Coffee   蓝山咖啡
wooden barrel   木桶

Step 2　Consolidation Training

Ⅰ　Complete the following sentences with the words in the box. Change the form if necessary.

> expensive　consuming　create　acceptable　transport

1. Firms have to be responsive to _____ demand.
2. She won't _____ advice from anyone.
3. The operating _____ is thought to be much higher.
4. The new museum must be accessible by public _____.
5. We want to _____ jobs for the unemployed.

Ⅱ　There are ten English words and terms numbered 1 through 10 in the left column. You should match them with their Chinese equivalents marked A through J in the right column and write down the corresponding letter in the right blank.

1. packaging	A. 真空包装
2. quality	B. 黄麻袋
3. jute bag	C. 包装
4. burlap-like	D. 减少
5. decline	E. 重点消费群体
6. vacuum packaging	F. 卖点
7. focus consuming groups	G. 麻布般的
8. superior quality	H. 可接受的
9. acceptable	I. 质量
10. selling point	J. 卓越的质量

1. _____　2. _____　3. _____　4. _____　5. _____　6. _____
7. _____　8. _____　9. _____　10. _____

Answers to the Tasks

Task Module 4　Extensive Reading

Learn More about the Processing of Coffee Fruit

After being picked from a coffea, the cherries could be processed in two main methods. They are a dry process and a wet process. The beans with the former process are often referred to as Natural Beans or Dry Beans while the latter one normally as Washed Beans.

For the dry process, the first step is to put all cherries through a quick wash. Then dry the cherries onto patios or raised beds for about two weeks. The cherries will lose their bright red colour to brown and shrink further. Then the dry natural cherries are ready for the next processing stage.

A wet process requires the use of considerable quantities of water and specific equipment. After immersion in water, the good ripe cherries will sink while the bad or unripe ones will float. Then the pulpers will help to remove the outer layers of cherries but leave the mucilage intact. Then the mucilage-covered beans will be separated according to their weight. The cherries will have a fermentation period from 12 to 72 hours till the mucilage breaks down and is washed off. The washed beans will be dried outside for 4-10 days, and then be sorted by hand with the aim of removing all damaged beans. The time will be shorter if the beans are machine-dried. After any treatment in the wet processes, the coffee green beans with reddish or brown patches are produced.

Task Module 5   Creative Project

Try to discuss this with your partner and find out the processing steps from coffee fruit to green coffee beans. Then draw the processing flow chart after your discussion. You are encouraged to collect more information from the library or through the internet if necessary.

Task 2　Roasting

 ## Part One　Task Introduction

1. Teaching Objectives

Knowledge Objectives	1. Understand the basic market demand for coffee beans. 2. Understand the roasting principle and roasting degree of coffee. 3. Learn the technical characteristics and operation steps of roasting coffee beans.
Skill Objectives	1. Be able to combine theory with practice, learn to identify the degree and characteristics of coffee roasting, and know how to identify and purchase roasted coffee beans. 2. In the process of learning English, feel the way of thinking in English, and improve the logic, speculation and innovation of thinking. 3. Be able to use English to understand and express information, views and feelings more accurately, and conduct effective cross-cultural communication.
Competency Objectives	1. Cultivate the sense of collaboration and teamwork. 2. Broaden the international vision, respect the diversity of the world culture, and build a sense of community with a shared future for mankind. 3. Build confidence and set up a correct view of English learning. 4. Learn to understand the key points of coffee baking technology and cultivate the craftsmanship spirit of excellence.

2. Teaching Focuses

(1) The roasting degree of coffee and the method of selecting beans.

(2) Baking technology of coffee beans.

3. Teaching Difficulty

Effective intercultural communication in English.

Part Two Task Implementation

Task Module 1 Pre-tasks

Step 1 Group Cooperation

Work in pairs. Discuss the professional terms in the box and get familiar with them. You can look them up online with your smartphones if necessary.

roast green coffee bean acid caffeine single-origin medium roast

Step 2 Panel Discussion

Work in pairs to discuss the following issues and exchange views.

1. Why do people roast coffee bean?

2. Which will you choose, the roasted coffee beans bought from a specialty café or the ones roasted by yourself at home? Why?

3. Do you know how to roast coffee beans?

Task Module 2 Reading and Training

Step 1 Reading

Coffee Beans Roasting

Roasting is a process that produces the characteristic flavor of coffee by changing the taste of green coffee beans, with the direct heating. It converts the chemical and physical properties of green beans into roasted beans. Though green beans include similar even higher levels of acids, protein, sugars and caffeine as those of baked beans, they are short of the characteristic taste of roasted ones which has been mainly created by the Maillard Reaction and Caramelization Reaction during roasting process.

The coffee market supplies unroasted beans and roasted beans to satisfy consumers' different demands. As green coffee beans are more stable than roasted ones, green beans are more welcome for import and export. According to the data of Alibaba.com, from January to September of 2020, China's imports of green coffee beans accounted for 51% of the total imports of the industry and exports for 89%. The vast majority of coffee is often baked commercially on a large scale, close to where it will be consumed. But many personalized cafés prefer to commercially roast the "single-origin" coffees by themselves on a small scale. And some coffee drinkers would treat "roast coffee at home" as a hobby which could help them to experiment with various beans and test different roasting methods. Meanwhile they could enjoy the fresh coffee as well as save their money since the home roasting coffees are cheaper than that of cafés.

Before roasting, the green coffee beans should be selected. All the defective beans that affect the flavor of coffee products should be picked out. Then the roasting could be started.

About the coffee roasting level, it could be roughly divided into Light Roast, Medium Roast and Deep Roast. According to Specialty Coffee Association (short for SCA), it could be subdivided into Light Roast, Cinnamon Roast, Medium Roast, High Roast, City Roast, Full City Roast, French Roast and Italian Roast. Then how to determine the degree of roast? For professional roasters, they will use a combination of temperature, smell, color and sound to monitor the roasting process. For example, when talking about evaluating the roast by the bean's color, it refers to observing the color change of coffee beans during roasting. As the beans absorb heat, The color first turns yellow, and then gradually turns dark to brown. In the later stages of roasting, oil will appear on the surface of beans. The beans will darken further until they are taken out from the heat source.

Word and Phrase Bank

New Words

direct /dəˈrekt/ *adj.*　直接的

chemical /ˈkemɪkl/ *adj.*　化学的

protein /ˈprəʊtiːn/ *n.*　蛋白（质）

during /ˈdjʊərɪŋ/ *prep.*　在……期间

unroasted /ʌnˈrəʊstɪd/ *adj.*　未经焙烧的

commercially /kəˈmɜːʃəlɪ/ *adv.*　商业上

defective /dɪˈfektɪv/ *adj.*　有问题的

subdivide /ˈsʌbdɪvaɪd/ *vt.*　再分，细分　*vi.*　细分，再分

combination /ˌkɒmbɪˈneɪʃn/ *n.*　结合（体）

observe /əbˈzɜːv/ *v.*　观察，看到

absorb /əbˈzɔːb/ *v.*　吸收

gradually /ˈɡrædʒuəli/ *adv.*　逐渐地

Phrases & Expressions

Maillard Reaction　美拉德反应

Caramelization Reaction　焦糖化反应

according to　根据

vast majority of　绝大多数

on a large scale　大规模的

single-origin　单一来源的

pick out　挑选出

Cinnamon Roast　轻中度烘焙（即肉桂烘焙）

City Roast　城市烘焙

Full City Roast　深城市烘焙

on the surface of　表面上

take out　取出

Step 2   After-reading Training

I   Answer the questions.

1. Which coffee beans are more welcome for import and export?

_____

2. How to classify roasting level according to Speciality Coffee Association?

_____

3. How to determine the degree of roast for professional roasters?

_____

4. What is the proportion of China's imports of green coffee beans and exports from January to September of 2020?

_____

5. What is roasting?

_____

II   Read the statements and tick the correct boxes.

Statements	Right	Wrong	Not mentioned
The coffee market only supplies roasted beans to satisfy consumers' different demands.			
About the coffee roasting level, it could be roughly divided into Light Roast, Medium Roast and Deep Roast.			
For professional roasters, they will use a combination of temperature, smell, color and sound to monitor the roasting process.			
In the later stages of roasting, oil will appear on the subsurface of beans.			
As the beans absorb heat, the color first turns yellow, and then gradually turns dark to brown.			

Answers to the Tasks

Task Module 3　Communication Tasks

Step 1　Role-playing

A: Look, our automatic coffee bean roaster finally arrived. There is also an electronic scale as a free gift.

B: Nice. Do you know how to use it?

A: No. But don't worry. We have the instructions. Its baking capacity is 150 g. Go and weigh 100 g coffee beans for me.

B: Here you are. The single production capacity of the coffee roasting machine in our training center is 3 kg, isn't it?

A: Yes. Compared to that machine, this one is quite small.

B: Indeed. How can we operate this one?

A: According to the instructions, the first step is to pour down the beans into this bin. Then move this sealing handle to the right.

B: OK. How about the next step?

A: Press the Power button. Then select the roasting degree you want with this Roasting Color button. Start the machine with this Roasting button. Now it bakes automatically.

B: Amazing. Here the temperature is on the screen.

A: Yes. If you want to stop the roasting, just press the Cancel button.

B: The operation is really simplified. How long will it take?

A: There are 8 roasting colors involving light roast, cinnamon roast, medium roast, high roast, city roast, full city roast, French roast and Italian roast. It depends on which color you've selected. From this transparent visible glass lid, it's easy to confirm the roasting state of coffee beans.

B: There is quite low noise. Oh, have you heard the cracking sound?

A: Yes. Just like popcorn popping. It's the first crack.

B: Is it smokeless?

A: Yes. This specially designed smoke eliminator can effectively absorb the harmful gas generated during baking. This function of safety and environmental protection is one of the reasons for me to select this product.

B: You've made the smart decision. Smokeless roasting is very important for our health.

A: It's done now. We move this sealing handle to the left. And take the bin out.

B: I love the flavor. Have you noticed the color? It seems like medium light brown. Right?

A: Yes. The roasted beans are in the big bin while the shells in the small bin. The separation design really saves time and effort.

B: You're absolutely right. Compared to that large and complicated roasting machine, this small automatic roaster is more proper for me.

Word and Phrase Bank

New Words

automatic /ˌɔːtəˈmætɪk/ *adj.*   自动的

scale /skeɪl/ *n.*   天平；秤

instruction /ɪnˈstrʌkʃn/ *n.*   指令；用法说明；操作指南

machine /məˈʃiːn/ *n.*   机器

capacity /kəˈpæsətɪ/ *n.*   容量

seal /ˈsiːlɪŋ/ *v.*   密封

temperature /ˈtemprətʃə(r)/ *n.*   温度

simplify /ˈsɪmplɪfaɪ/ *v.*   使（某事物）简单

cracking /ˈkrækɪŋ/ *n.*   爆裂声；噼啪声

popcorn /ˈpɒpkɔːn/ *n.*   爆米花

eliminator /ɪˈlɪmɪneɪtə/ *n.*   消除器

Phrases & Expressions

compared to   与……相比

pour down   倾泻

depend on   依靠

environmental protection   环境保护

Step 2   Consolidation Training

I   Match the words on the left with their meanings on the right.

1. machine	A. 有关环境的
2. visible	B. 温度
3. temperature	C. 决定
4. environmental	D. 环境保护
5. instruction	E. 机器
6. electronic scale	F. 依靠
7. decision	G. 说明
8. environmental protection	H. 无烟烘焙
9. depend on	I. 看得见的
10. smokeless roasting	J. 电子秤

1. _____   2. _____   3. _____   4. _____   5. _____   6. _____   7. _____

8. _____   9. _____   10. _____

Ⅱ Decide whether the following statements are true (T) or false (F).

(　　) 1. There are 8 roasting colors involving light roast, cinnamon roast, medium roast, high roast, city roast, full city roast, French roast and Italian roast.

(　　) 2. The single production capacity of the coffee roasting machine in our training center is 5 kg.

(　　) 3. This specially designed smoke eliminator can effectively absorb the harmful gas generated during baking.

(　　) 4. From this transparent visible glass lid, it' s hard to confirm the roasting state of coffee beans.

(　　) 5. According to the instructions, the first step is to pour down the beans into this bin. Then move this sealing handle to the right.

Answers to the Tasks

Step 1　Role-playing

A: John, I want to try and buy some roasted coffee bean. Do you have any suggestions?

B: Yes. As the main enemies of roasted beans are oxygen, moisture, light, heat and strong odors, try to choose beans protected in containers with lids or a one-way air valve. For example, opaque airtight bag with a one-way valve is a good package.

A: Some coffee beans are packed in kraft paper bags or vacuum-sealed bags.

B: In my opinion, kraft paper bags just could offer minimal protection for coffee. And the method of vacuum sealing removes excessive aroma volatiles from coffee, which has a negative impact on its sensory characteristics. I would not choose any of them.

A: I see. Anything else?

B: Remember to see the roast date. Some experts believe that beans roasted for more than two weeks are too old. I' m not sure about it. But it' s always right to buy as fresh as possible.

A: OK. I got it.

B: What flavor do you like?

A: I' m not sure.

B: In this case, try to get more information about the product. You can obtain it from the packaging and the barista. Select several types and buy a little for comparison and experiment until you find the quality and flavor you want.

A: What information will be on the packaging?

B: Ideally you can find the name of the producer and/or farm owner, expected flavor, how the coffee being processed, altitude, roast level, roasted and packed date and provenance which would tell you the species and/or variety the coffee is, whether it is a blend or single source, and the like.

A: It' s too complicated. Could I just select the expensive beans?

B: I'm afraid it's not a better choice. How about asking the baristas for recommendations and guidance? They're quite professional about bean selection and equipment usage.

A: Great. Would you like to go with me?

B: I'm sorry I have my hands full. How about tomorrow?

A: No problem.

Word and Phrase Bank

New Words

moisture /ˈmɔɪstʃə(r)/ *n.* 水分

opaque /əʊˈpeɪk/ *adj.* 不透明的

kraft /krɑːft/ *n.* 牛皮纸

minimal /ˈmɪnɪməl/ *adj.* 最小的，极小的

excessive /ɪkˈsesɪv/ *adj.* 过多的

volatiles /ˈvɒlətaɪl/ *n.* 挥发物

barista /bəˈriːstə/ *n.* 咖啡师

provenance /ˈprɒvənəns/ *n.* 出处

complicated /ˈkɒmplɪkeɪtɪd/ *adj.* 复杂难懂的

choice /tʃɔɪs/ *n.* 选择

guidance /ˈgaɪdns/ *n.* 指导

Phrases & Expressions

vacuum-sealed bag 真空密封袋

farm owner 农场主

expected flavor 预料中的味道

roast level 烘焙程度

single source 单一来源

equipment usage 设备使用

Step 2 Consolidation Training

I Complete the following sentences with the words in the box. Change the form if necessary.

main believe usage completely better

1. Some experts _____ that beans roasted for more than two weeks are too old.

2. I'm afraid it's not a _____ choice.

3. As the _____ enemies of roasted beans are oxygen, moisture, light, heat and strong

odors, try to choose beans protected in containers with lids or sneeze guards.

4. Coffee from vacuum-sealed bags or bricks often has been _____ degassed and gone stale before packaging.

5. They're quite professional about bean selection and equipment _____.

Ⅱ   There are ten English words and terms numbered 1 through 10 in the left column. You should match them with their Chinese equivalents marked A through J in the right column and write down the corresponding letter in the right blank.

1. vacuum-sealed bag	A. 选择
2. farm owner	B. 单一来源
3. expected flavor	C. 指导
4. roast level	D. 真空密封袋
5. single source	E. 最小的，极小的
6. equipment usage	F. 农场主
7. minimal	G. 设备使用
8. complicated	H. 烘焙程度
9. choice	I. 预料中的味道
10. guidance	J. 复杂难懂的

1. _____   2. _____   3. _____   4. _____   5. _____   6. _____

7. _____   8. _____   9. _____   10. _____

Answers to the Tasks

Task Module 4   Extensive Reading

Learn More about Roasting Green Coffee Beans

During the stages of roasting green coffee beans, cracking sounds could be heard from the roaster. When the temperature reaches about 196 ℃ (385 °F), the force of the steam eventually ruptures the cell structure of the beans, which emit the first cracking sound. The beans increase in size, gain an even color as cinnamon in appearance and begin to smell like coffee. The grassy odor has been removed but the toasted grain smell is noticeable. The aroma smells good and the acidity is sharp though the sweetness is still underdeveloped. The beans enter into a very light roast level which commonly called the stage of cinnamon roast. It is normally used in American coffee.

When the temperature is close to 224 ℃ (435 °F), the beans make a second crack caused by gas pressure. The beans are fully developed into brittle ones. The color becomes mediumdark brown. Tiny droplets or faint patches of oil appear on the surface of the beans. Roast character is outstanding. Bitterness is stronger than acidity. The medium dark brown level is normally called the stage of full city roast.

The different stages of roasting green beans create different characteristics of coffee beans. Besides the two stages mentioned above, the characteristics of the beans in other six stages are as follows:

No.	Common Roast Stages	Characteristics	Roasting Level	Flavor
1	Light Roast	The lightest baking degree. Yellowing phase. As the taste and aroma are insufficient, beans of this stage can hardly be used for tasting but usually used for inspection.	Light Roast	Lighter-body. Higher acidity, very little bitterness and no obvious roast taste. Mainly in floral and fruit flavors. Full origin character of the coffee is dominant.
2	Cinnamon Roast	The beans enter into a very light roast level. The first cracking sound starts from the roaster. The beans gain an even color as cinnamon in appearance. The acidity is sharp as the Maillard Reaction is not given full play.		
3	Medium Roast	Moderate light brown color with mottles in appearance. Bright acidity, moderate aroma, and a little bitter taste.	Medium Roast	As sugars have been further caramelized, other components have relatively been muted. The acidity and sweetness have been highlighted and little roast flavor has been developed. It is especially suitable for making single coffee.
4	High Roast	Moderate brown color. Bitterness and acidity is relatively balanced with good aroma and flavor but origin character is still preserved.		
5	City Roast	Brown color. Good balance of bitterness and acidity. Origin character and baking character coexist harmoniously.		

Continuation table

No.	Common Roast Stages	Characteristics	Roasting Level	Flavor
6	Full City Roast	A second cracking starts. Moderate-dark brown color with tiny faint patches on the surface. Roast character is obvious. Bitterness is stronger than acidity.	Deep Roast	Shiny and thin. Sugar has mostly burnt out. Little acidity left. Aromas and flavors of roast become clearly evident.
7	French Roast	Dark brown color. The surface is shiny with oil. Acidity and almost all inherent aroma or flavors of the coffee beans diminished while bitterness is very strong. Roast character is prominent.		
8	Italian Roast	Nearly black. Shiny on the surface with a thin body. The original flavour of the coffee beans disappeared while burnt tones become more distinct. Almost only roast, ashy and bitter tastes without acidity.		

Notice: According to the research, caffeine remained stable below 200 ℃ (392 °F) and completely decomposed at about 285 ℃ (545 °F). Given that the roasting temperature normally does not exceed 200 ℃ (392 °F) for a long time and rarely reaches 285 ℃ (545 °F), the caffeine content in coffee beans may not change much during roasting.

Task Module 5　Creative Task

　　In many specialty cafés, they will store various roasted coffee beans to get ready for offering single coffee or blended coffee to customers. Work in pairs. Try to introduce different roasting levels as well as the common roasting names to others to help them know more about the characteristics of roasting techniques. You are encouraged to collect more information from the library or through the internet if necessary.

Task 3   Grinding

 Part One   Task Introduction

1. Teaching Objectives

Knowledge Objectives	1. Understand different grinding methods of coffee beans. 2. Understand the differences between manual grinder and automatic grinder, and learn the basic operation process of the grinders. 3. Understand the differences between different coffee grounds and the applicable brewing method for each ground.
Skill Objectives	1. Combine theory with practice. Get familiar with the basic operation process through coffee bean grinding training, and study and draw the operation flow chart of different grinding methods. 2. Learn to distinguish the coarseness of coffee grounds, and design the matching diagram of coffee grounds and brewing methods. 3. In the process of learning English, feel the way of thinking in English, and improve the logic, speculation and innovation of thinking. 4. Be able to use English to understand and express information, views and feelings more accurately, and conduct effective cross-cultural communication.
Competency Objectives	1. Cultivate the sense of collaboration and teamwork. 2. Broaden the international vision, respect the diversity of the world culture, and build a sense of community with a shared future for mankind. 3. Through the training of distinguishing and contrasting ability, gain a sense of identity, improve self-confidence, and promote the development of a correct concept of English communication. 4. Combine practice with training. Strengthen training from different angles, and build design and innovative thinking. 5. Understand the key points of coffee bean grinding and the influence of coffee ground coarseness on the quality of coffee drinks, and cultivate the craftsmanship spirit of keeping improving and paying attention to quality.

2. Teaching Focuses

(1) Understand the basic operation process of coffee grinding.

(2) Compare the coffee grinding methods to understand the relationship between grinding degree and brewing methods.

3. Teaching Difficulty

Effective intercultural communication in English.

Part Two   Task Implementation

Task Module 1   Pre-tasks

Step 1   Group Cooperation

Work in pairs. Discuss the professional terms in the box and get familiar with them. You can look them up online with your smartphones if necessary.

> coffee grinder   manual grinder   electric grinder   grinding degree
> uniformity of coarseness

Step 2   Panel Discussion

Work in pairs to discuss the following issues and exchange views.

1. What kinds of coffee grinders have you ever seen or used?

2. What are the differences between a manual coffee grinder and an electric grinder?

3. What factors will you consider when choosing a coffee grinder?

Task Module 2   Reading and Training

Step 1   Reading

Grinding

If you want to improve the quality of the coffee you make and to achieve the correct texture, there is one very simple way for you to do it. That is to grind the fresh coffee beans with a good grinder. Basically there are two types of grinders in the market. One is manual grinder or hand grinder. The other is electric grinder. Compared to the electric one, the manual grinder is compact, lightweight, portable and easier to be cleaned. With the continuous innovation of technology, especially the application of the adjustable burrs, the hand-cranked manual grinder becomes more convenient to adjust the uniformity of coarseness. The hand grinder could make the coffee particle size more uniform. And no overheating during grinding is very helpful to maintain the original flavor and aroma of coffee. Though it usually has only 1 or 2 powder storage jars, it's favored and applied by people when at home, at the office, traveling or camping. You can slowly grind coffee while water boils. Then brew a cup of rich and smooth coffee for a nice and relaxing morning. Many people really enjoy the process. The manual grind starts to have its own market share.

However, although the electric grinders are more expensive than the manual ones, they still occupy the main market as they could grind more coffee easily and quickly. The friction and waste heat produced by the motor may heat the ground slightly which will influence the quality of the coffee. But the development of technology makes this impact smaller and smaller. In today's market, you can purchase electric grinders of different capacities for hotel, garage, household, coffee house or office. No matter what type you want, you could always buy one you like.

So how to choose a coffee grinder? Many factors should be considered, including the price, capacity, size, grinding degree, uniformity of coarseness, materials, noise production, burr or blade, and so on. And the grinding degree and uniformity of coarseness are regarded as the two most important elements for improving the quality of the coffee.

Word and Phrase Bank

New Words

achieve /ə'tʃiːv/ *v.* 达到，实现

texture /'tekstʃə(r)/ *n.* 质地，口感

grind /graɪnd/ *v.* 磨碎，把……磨成粉

grinder /'graɪndə(r)/ *n.* 研磨机

manual /'mænjuəl/ *adj.* 手工的

electric /ɪ'lektrɪk/ *adj.* 电的

compact /'kɒmpækt/ *adj.* 小型的，袖珍的

lightweight /'laɪtweɪt/ *adj.* 轻的

portable /'pɔːtəb(ə)l/ *adj.* 便携式的，轻便的

continuous /kən'tɪnjuəs/ *adj.* 连续不断的，持续的

innovation /ˌɪnə'veɪʃ(ə)n/ *n.* 革新，创新

adjustable /ə'dʒʌstəb(ə)l/ *adj.* 可调节的

burr /bɜː(r)/ *n.* 刀盘，磨盘

uniformity /ˌjuːnɪ'fɔːməti/ *n.* 统一，一致

coarseness /'kɔːsnəs/ *n.* 粗，粗糙

particle /'pɑːtɪk(ə)l/ *n.* 微粒，颗粒

jar /dʒɑː(r)/ *n.* 玻璃罐

overheat /ˌəʊvə'hiːt/ *vt. & vi.* 过热

occupy /'ɒkjupaɪ/ *v.* 占领

slightly /'slaɪtlɪ/ *adv.* 稍微，轻微地

ground /graʊnd/ *n.* 粉末，渣滓

element /'elɪmənt/ *n.* 要素

Phrases & Expressions

manual grinder　手动磨豆机，手动研磨机

electric grinder　电动磨豆机，电动研磨机

hand-cranked manual grinder　曲柄手动研磨机

market share　市场占有率

grinding degree　研磨度

uniformity of coarseness　研磨粗细均匀度

Step 2　After-reading Training

I　Answer the questions.

1. What process do people enjoy when they use a hand grinder?

_____

2. Why do electric grinders occupy the main market even though they are more expensive than the manual ones?

_____

3. Why the quality of coffee would be influenced when using electric grinders?

_____

4. What factors should be considered when choosing a coffee grinder?

_____

5. What are the two most important factors for improving the quality of the coffee?

_____

II　Read the statements and tick the correct boxes.

Statements	Right	Wrong	Not mentioned
Compared to the manual grinder, the electric grinder is compact, lightweight, portable and easier to be cleaned.			
The manual grinder is favored by people at home because it helps to maintain the rich and original flavor and aroma of coffee.			
The electric grinder is more expensive than the manual one because it is larger and more complicated.			
Development of technology eliminates the negative influence on the quality of the coffee when using an electric grinder.			

Continuation table

Statements	Right	Wrong	Not mentioned
The grinding degree and uniformity of coarseness are regarded as the two most important elements for improving the quality of the coffee.	√		

Task Module 3　Communication Tasks

Situation 1

Answers to the Tasks

Step 1　Role-playing

A: Kayla, how to use the coffee grinder?

B: OK. Which one do you want to use?

A: What are the differences between them?

B: They are both electric grinders. This is filter style. That one is espresso style.

A: Could you tell me more in detail?

B: Yes. The grinding degree of this filter-style grinder can be adjustable. It's quick and convenient to use but can't grind fine or super-fine ground.

A: How about this one? It looks quite modern.

B: It's the expresso-style grinder. We bought it last year as we often need to grind large quantities of coffee for expresso.

A: Tell me how to use this one.

B: OK. This is hopper for storing beans. This is the scale adjustment which controls the exact particle size. It has a digital timer function. The figure on the screen is the grinding time. You can adjust it with the following small buttons. The orange button controls the on/off for the whole machine. The black button is for the starting of the grinding. Now let me show you how to run it. Pass me the beans.

A: Here you are.

B: The first step is to fill 20 g beans into the hopper. The next step is to turn it on with the orange button. The screen lights up. Then you can move the scale circle to make the adjustment of the particle size if necessary. Now it points at 4. Would you like to adjust it to 8 for me?

A: Let me have a try. Is it correct?

B: Yes, very good. Then adjust the time. For example, we set 6 seconds with this button. Now press the black button to start to grind.

A: Wow, it works.

B: The machine will automatically stop when the time is up.

A: Interesting. Could I grind some coffee beans by myself?

B: Sure, go ahead.

Word and Phrase Bank

New Words

convenient /kənˈviːniənt/ *adj.* 方便的，便利的

fine /faɪn/ *adj.* 小颗粒的，精细的

quantity /ˈkwɒntəti/ *n.* 数量，大量

hopper /ˈhɒpə(r)/ *n.* 豆仓

scale /skeɪl/ *n.* 标度，刻度

timer /ˈtaɪmə(r)/ *n.* 定时器，计时器

function /ˈfʌŋkʃ(ə)n/ *n.* 功能

particle /ˈpɑːtɪk(ə)l/ *n.* 颗粒

automatically /ˌɔːtəˈmætɪkli/ *adv.* 自动地

Phrases & Expressions

in detail 详细地

scale adjustment 刻度调节器

grinding time 研磨时间

turn on 开机

particle size 颗粒大小

Step 2 Consolidation Training

Ⅰ Match the words on the left with their meanings on the right.

1. automatically	A. capable of being changed so as to match or fit
2. function	B. funnel-shaped receptacle; contents pass by gravity into a receptacle below
3. quantity	C. to perform a task without needing to be constantly operated by a person
4. particle	D. an indicator having a graduated sequence of marks
5. convenient	E. suited to your comfort or purpose or needs
6. scale	F. what something is used for
7. hopper	G. marked by strict and particular and complete accordance with fact
8. timer	H. how much there is of something that you can quantify
9. adjustable	I. a tiny piece of anything
10. excact	J. a timepiece that measures a time interval and signals its end

1. _____ 2. _____ 3. _____ 4. _____ 5. _____ 6. _____ 7. _____

8. _____ 9. _____ 10. _____

II   Decide whether the following statements are true (T) or false (F).

(   ) 1. The filter-style grinder is quick and convenient to use and can grind super-fine or fine ground.

(   ) 2. The orange button can turn the expresso grinder machine on or off.

(   ) 3. The expresso grinder can grind coffee beans into different particle sizes.

(   ) 4. One needs to press the black button to end the working process of the espresso machine.

(   ) 5. The scale circle adjusts the grinding time length.

Answers to the Tasks

Situation 2

Step 1   Role-playing

A: Kayla, why do we need to adjust the scale and the time?

B: Let's do an experiment. We'll grind the same amount of beans on three different scales but at the same time. After that we can make a comparison of the grounds.

A: OK. I'm eager to see the results.

B: Now the first time is to grind at the scale of 4 in 6 seconds. Now done. Pour the grounds in cup A.

A: OK. Let me mark the cup as A.

B: This is to grind at the scale of 8 in 6 seconds.

A: Let me pour the grounds into cup B.

B: Nice. Now to grind at the scale of 12 in 6 seconds.

A: Pour the final grounds into cup C.

B: What are the differences between them?

A: The grounds in cup A has the smallest particle size, while cup B has the medium size and cup C has the largest size.

B: Yes. We call the grounds in cup A fine grind, cup B medium grind and cup C coarse grind. The results will be like this if the beans of the same amount are ground at the same scale but at different times. The less time you use, the coarser of the coffee powder will be and vice versa.

A: I see.

B: Once we've finished the grinding, we need to clean the grinder. Would you like to clean it with me?

A: I'd love to.

 **Word and Phrase Bank**

New Words

experiment /ɪkˈsperɪmənt/ *n.* 实验

amount /əˈmaʊnt / *n.* 数量

comparison /kəmˈpærɪs(ə)n/ *n.* 比较，对照

pour /pɔː(r) / *v.* 倒出

medium /ˈmiːdiəm/ *adj.* 中等的，中号的

ground /graʊnd/ *n.* 碎末，粉末

coarse /kɔːs/ *adj.* 粗糙的

powder /ˈpaʊdər/ *n.* 粉，粉末

Phrases & Expressions

medium size 中等尺寸

fine grind 细粉

medium grind 中度粉

coarse grind 粗粉

vice versa 反之亦然

Step 2 Consolidation Training

Ⅰ Complete the following sentences with the words in the box. Change the form if necessary.

> particle difference comparison grind eager experiment scale pour
> coarse vice versa

1. What's the _____ between an espresso grinder and a filter grinder?

2. Many people do not like the idea of _____ on animals.

3. Grounds in cup A has the smallest _____ size, while cup B has the medium size.

4. How much does it read on the _____?

5. The second half of the game was dull by _____ with the first.

6. The old man opens the gift with his _____ hands.

7. She _____ her teeth when she is asleep.

8. _____ the sauce over the pasta.

9. When one side wins, the other loses, and _____.

10. He was _____ to communicate his ideas to the group.

II   There are ten English words and terms numbered 1 through 10 in the left column. You should match them with their Chinese equivalents marked A through J in the right column and write down the corresponding letter in the right blank.

1. adjust	A. 反之亦然
2. scale	B. 比较
3. ground	C. 粗糙的
4. comparison	D. 调整
5. mark	E. 碎末，粉末
6. fine grind	F. 中度粉
7. medium size	G. 粉，粉末
8. coarse	H. 刻度
9. powder	I. 标注
10. vice versa	J. 细粉

1. _____   2. _____   3. _____   4. _____   5. _____   6. _____
7. _____   8. _____   9. _____   10. _____

Task Module 4   Extensive Reading

Answers to the Tasks

Learn More about the Grinding of Coffee

Coffee beans may be ground in various ways no matter by machine or by hand. The machine with a burr is called a burr grinder which uses revolving elements to shear the seed. And the machine with a blade is named a blade grinder which cuts the seeds with blades moving at high speed. For most brewing methods a burr grinder is regarded superior and more welcomed as it produces particles of a uniform size while the blade one often chops coffee beans into grounds with inconsistent size. And coffee grounds can also be made by a mortar with a pestle or a hammer. The cowboys in old days would use a hammer to crush the seeds and then boiled the coffee with a pot over a campfire.

Do you know which grind matches coffee brewing method better? According to the particle size of coffee powder, there are four kinds of grinds. They are the super-fine grind (also named as Turkish grind), fine grind, medium-coarse grind and coarse grind. Some connoisseurs suggest that the super-fine grind creates finest powdery grounds. It is better to use Ibrik coffee pot to extract the maximum flavor of the powder. Fine grind produces fine grounds, so it is very suitable for espresso machine which will extract in a balanced shot. Medium-coarse grind gives

you the grounds of medium particle size which is popular for many brewing methods, such as filter pour-over, stove-top pot, French press, syphon, etc. And coarse grind offers you the coarsest ground which is more proper for the brewers using equipment like flannel drip or drip coffee machine. With this equipment, the water in it could penetrate the cell structure of coarse grounds. The pleasant solubles would be dissolved and excessive bitterness could be avoided through this process.

Task Module 5   Creative Project

Regular cleaning of the coffee grinder is a key to effective protection of the equipment. How to clean it? Try to clean a grinder with your partners. And write down the procedures. Discuss your procedures with your partners before finalizing them. You are encouraged to collect more information from the library or through the internet if necessary.

项目 2

咖啡加工

任务1　咖啡果的处理

 第1部分　任务介绍

咖啡果的处理

1. 教学目标

知识目标	1. 了解咖啡果的采摘和常见的处理方法。 2. 了解咖啡生豆的包装特点。 3. 了解咖啡生豆的市场需求。
技能目标	1. 理论联系实际，学习绘制咖啡果的处理流程图。 2. 在学习英语的过程中，感受英语思维方式，提升思维的逻辑性、思辨性和创新性。 3. 能够运用英语，比较准确地理解和表达信息、观点、情感，能有效地进行跨文化交际与沟通。
素质目标	1. 培养与人协作的精神和团队合作的意识。 2. 拓宽国际视野，尊重多元文化，树立人类命运共同体意识。 3. 树立信心，积极适应环境需求，学会及时调整心态，树立正确的英语学习观。 4. 训练学生的逻辑思维能力和信息加工能力，培养学生的创新意识。

2. 教学重点

（1）咖啡果的处理。

（2）咖啡生豆包装的作用与影响。

3. 教学难点

能用英语进行跨文化交际与沟通。

 第2部分　任务实施

任务模块1　前置任务

步骤1　小组合作

两人一组，讨论下列方框中的专业用语并加以熟悉。如果需要，可以使用智能手机在线查找。

> cherry berry harvest processing labor-intensive density mass consumer

步骤 2 小组讨论

两人一组，讨论以下问题并交流观点。

1. 你知道如何采摘咖啡果吗？
2. 你知道咖啡果的处理有几个步骤？分别是哪些步骤？

任务模块2 阅读与训练

咖啡果的处理

为了变成咖啡豆，需要加工咖啡树的果实。咖啡树的果实通常被称为咖啡樱桃或浆果。但是，首先要从树上采摘咖啡果。

一般来说，收集咖啡果有两种方法。第一种方法是成串剥离。无论成熟状态如何，所有咖啡果都用机械剥离器从树上采下来。这种方法通常用于巴西和澳大利亚的大中型农场，那里气候干燥，咖啡田广阔。第二种方法是手工采摘。在大多数国家，只有成熟度最高的果实是手工采摘的。由于这种采摘方式是劳动密集型的，成本更高，因此主要用于收获高质量的阿拉比卡豆。

采摘后，所有的咖啡果应在几个小时内进行处理，以保证其质量。在用干磨机进行处理之前，有两种主要的方法处理这些咖啡果，包括自然日晒处理法和水洗处理法。通过这些方法，咖啡果将变成外壳收缩的棕色干咖啡豆或仍包裹着一层内果皮的咖啡豆。之后将这些咖啡豆转移到干磨机中，有助于去除它们的外层果肉和内果皮。咖啡豆的分拣可以由干磨机或手工完成。

咖啡生豆可按重量、大小和/或颜色来分类。低质量或中等质量的咖啡生豆将被倒入容器中，以生产更为便宜的大众消费咖啡。优质的咖啡生豆将被运往专业市场。传统的黄麻袋或聚乙烯编织袋具有极强的多孔性，容易使咖啡生豆暴露在周围环境中。这种不良的储存方式可能会使咖啡生豆产生粗麻布般的味道，其积极品质可能会消失。为了保持咖啡生豆的品质，近年来，精品咖啡市场开始选择一些增强存储性能的方法，例如，将咖啡生豆存储在有内衬塑料的保护袋中或真空包装中。真空包装可能会进一步降低咖啡生豆在大气湿度下受氧气影响的程度，但它的成本更高。

对于咖啡来说，从农场到烘焙机是一段漫长的旅程。简单来说，它包含种植、采摘、筛选、处理和运输。完成上述步骤后，咖啡豆的购买者先对咖啡生豆进行烘焙、研磨和冲泡，然后为他人提供咖啡饮品。

任务模块3　场景交际任务

场景1

A：克里斯蒂娜，我对咖啡果实的处理有些困惑。你能帮我吗？

B：我试试看。你的问题是什么？

A：谁来对咖啡果实进行干磨？是农民还是其他人？

B：据我所知，一些拥有自己的工厂的农民会自己加工咖啡果实，这样他们就可以在咖啡生豆出口前保持对咖啡生豆品质的控制。而其他没有自己工厂的生产商可能会将咖啡果实出售并送去处理厂，在那里进行果实的干燥和碾磨。

A：我明白了。干磨处理后，将对咖啡生豆做些什么？

B：它们将被分为不同的类别，以呈现咖啡豆的质量。人们常说，在咖啡的世界里，任何咖啡豆，从最便宜的到排名前1%，都有买家。

A：为什么？

B：低质量的咖啡生豆将以最低的价格出售，以生产更便宜的大众消费者咖啡。它们的特点是有令人不快的苦味、涩味以及刺激性强的气味。

A：最好的呢？

B：质量最好的咖啡豆会更贵，因为它们更芳香，更顺滑，更醇厚，芳香油含量更高，但有机酸含量更低。

A：正如我们经常说的，质量决定价格。

场景2

A：特蕾莎，你为什么想改变编号为225的咖啡豆的包装？

B：因为它有最高的品质，编号225是我们最贵的产品之一。如果我们使用黄麻袋，这种糟糕的包装只会产生类似粗麻布的味道，并降低其品质。那样的话，我们如何以更高的价格出售它呢？

A：但是，如果你把黄麻袋改成真空包装，成本会立即增加。

B：别担心。我们已经调查了几个重点消费群体。结果表明，他们更喜欢更好的包装，以获得品质优良的咖啡豆。价格上涨是可以接受的。

A：我还是担心价格问题。价格上涨的百分比是多少才能被接受？

B：我们来进行一次产品测试怎么样？同时，我们还可以尝试用我们的新包装创建我们的新品牌。

A：好主意。你想怎么做呢？

B：我正在考虑蓝山咖啡的包装方法。

A：你的意思是？

B：与大多数咖啡生豆用60千克/袋的布袋包装和运输不同，蓝山咖啡用70千克/桶的木桶储存、运输和销售。众所周知，牙买加是最后一个仍然用传统木桶运输咖啡的国家。当我们从牙买加购买蓝山咖啡豆时，包装是他们的卖点之一。

A：是的，我知道你现在在想什么了。

任务模块4　知识拓展

了解更多关于咖啡果加工的信息

从咖啡树上采摘下来后，咖啡果可以用两种主要方法进行处理，即干燥法和水洗法。采用前一种方法处理过的咖啡豆通常称为自然豆或干燥豆，而采用后一种方法处理过的咖啡豆通常称为水洗豆。

对于干燥法，第一步是先将所有的果实快速清洗，然后将果实放在露台或晒架上干燥大约两周。果实会从鲜红色变为棕色，并且会萎缩。之后，干燥的果实准备进入下一个处理阶段。

水洗法需要使用大量的水和专用设备。在浸入水中后，好的熟果会下沉，而坏的或未成熟的果实会浮起来。接着，使用剥皮机将有助于去除果实的果肉，但保持黏膜完整。然后，黏膜包裹着的咖啡豆将根据重量进行分离。果实将有12～72小时的发酵期，直到黏膜分解并洗掉。水洗后的豆子将在室外干燥4～10天，然后手动分拣去除所有受损的豆子。如果用机器干燥豆子，时间会更短。水洗处理后的咖啡生豆会产生红色或棕色斑点。

任务模块5　创意任务

试着与你的组员讨论，找出从咖啡果到咖啡生豆的加工处理步骤，并绘制出其处理流程图。如有必要，鼓励你从图书馆或通过互联网收集更多信息。

任务模块6　任务执行能力考核

序号	考核细分项目	标准分数（分）	得分（分）
1	专业用语	10	
2	任务资料的基础训练	20	
3	场景任务1专项训练	20	
4	场景任务2专项训练	20	
5	知识拓展	10	
6	创意任务	20	

任务模块7 目标达成考核

评分考核项目	标准分数（分）	个人自评	小组互评	教师评分
理论知识	30			
技能训练	40			
职业素养	30			
总分（分）	100			
综合总分				
说明	综合总分=个人自评（占总分的20%）+小组互评（占总分的20%）+教师评分（占总分的60%）			

任务2　咖啡豆烘焙

 第1部分　任务介绍

咖啡豆烘焙

1. 教学目标

知识目标	1. 了解咖啡豆市场的基本需求。 2. 了解咖啡豆的烘焙原理和烘焙程度。 3. 学习咖啡豆烘焙的技术特点和操作步骤。
技能目标	1. 能够理论联系实践，学会辨别咖啡烘焙程度和特点，能够对烘焙好的咖啡豆进行辨别与选购。 2. 在学习英语的过程中，感受英语思维方式，提升思维的逻辑性、思辨性和创新性。 3. 能够运用英语，比较准确地理解和表达信息、观点、情感，能够有效地进行跨文化交际与沟通。
素质目标	1. 培养与人协作的精神和团队合作的意识。 2. 拓宽国际视野，尊重多元文化，树立人类命运共同体意识。 3. 树立信心，不畏困难，积极探索，树立正确的英语学习观。 4. 学习领会咖啡烘焙技术关键点，培养精益求精的工匠精神。

2. 教学重点

（1）咖啡的烘焙程度和选豆方法。

（2）咖啡豆的烘焙技术。

3. 教学难点

能用英语进行跨文化交际与沟通。

 第2部分　任务实施

任务模块1　前置任务

步骤1　小组合作

两人一组，讨论下列方框中的专业用语并加以熟悉。如果需要，可以使用智能手机

在线查找。

> roast   green coffee bean   acid   caffeine   single-origin   medium roast

步骤 2   小组讨论

两人一组，讨论以下问题并交流观点。

1. 为什么人们要烘焙咖啡豆？

2. 是从专业咖啡馆购买的已经烘焙好的咖啡豆还是在家自己烘焙咖啡豆？你会选择哪一种？为什么？

3. 你知道如何烘焙咖啡豆吗？

任务模块2   阅读与训练

咖啡豆烘焙

烘焙是通过直接加热来改变咖啡生豆味道，从而产生咖啡特有风味的过程。这个过程改变了咖啡生豆的化学成分和物理特性，将其烘焙成咖啡豆。虽然咖啡生豆含有与烘焙后的咖啡豆相似甚至更高水平的酸、蛋白质、糖和咖啡因，但它们缺乏后者的特殊味道，这是烘焙过程中美拉德反应和焦糖化反应产生的。

咖啡市场供应咖啡生豆和烘焙后的咖啡豆，以满足消费者的不同需求。由于咖啡生豆比烘焙后的咖啡豆更稳定，因此咖啡生豆更受进出口贸易的欢迎。根据阿里巴巴网站的数据，2020年1—9月，中国咖啡生豆进口占行业进口总额的51%，出口占89%。在商业领域，绝大多数咖啡通常会在离消费地很近的地方进行大规模烘焙。但许多个性化咖啡馆更喜欢小规模自行烘焙单品咖啡。一些咖啡饮用者会把"在家里烘焙咖啡"当作一种爱好，这样可以帮助他们试验各种咖啡豆和测试不同的烘焙方法。同时，他们可以享受新鲜美味的咖啡，也可以省钱，因为家庭烘焙咖啡比咖啡馆更便宜。

烘焙前，应对咖啡生豆进行挑选。所有影响咖啡饮品风味的有缺陷的咖啡生豆都应挑选出来，然后才开始烘焙。

关于咖啡烘焙程度，大致可分为轻度烘焙、中度烘焙和深度烘焙。根据精品咖啡协会（简称SCA）的定义，它可以分为轻度烘焙、轻中度烘焙（即肉桂烘焙）、中度烘焙、深度烘焙、城市烘焙、深城市烘焙、法式烘焙和意式烘焙。那么，如何确定烘焙的程度呢？对于专业烘焙师，他们将使用温度、气味、颜色和声音的组合来监控烘焙过程。例如，当说到通过咖啡豆的颜色来评价烘焙时，它指的是观察咖啡豆在烘焙过程中的颜色变化。当咖啡豆吸收热量时，首先颜色变黄，然后逐渐变暗为棕色。在烘焙的后期，咖啡豆表面会出现油。咖啡豆将进一步变黑，直到从热源中取出。

任务模块3 场景交际任务

场景1

A：看，我们的自动咖啡豆烘焙机终于来了。还有一个电子秤作为赠品。

B：太好了。你知道怎么用吗？

A：不知道，不过不用担心，我们有说明书。它的烘焙重量是150克。请给我称100克咖啡豆。

B：给你。我们培训中心咖啡烘焙机的烘焙重量是3千克，是吗？

A：是的。与那台机器相比，这台机器很小。

B：确实如此。我们怎么操作这台机器？

A：根据说明，第一步把豆子倒进这个箱子。然后向右移动这个密封手柄，盖上盖子。

B：好，下一步怎么做？

A：按下"电源"按钮。然后用这个烘焙颜色按钮选择所需的烘焙程度。用这个"烘焙"按钮可以启动机器。现在它开始自动烘焙了。

B：太棒了。这个屏幕会显示温度。

A：是的。如果你想停止烘焙，只需按"取消"按钮。

B：操作非常简单。需要多长时间？

A：有8种烘焙颜色可以选择，包括轻度烘焙、轻中度烘焙、中度烘焙、深度烘焙、城市烘焙、深城市烘焙、法式烘焙和意式烘焙。烘焙时间取决于你选择的颜色。从这个透明的玻璃盖上可以很容易确认咖啡豆的烘焙状态。

B：噪声很小。哦，你听到爆裂声了吗？

A：听到了。就像爆米花一样。这是第一次爆裂。

B：它是无烟的吗？

A：是的。这种特殊设计的消烟器能有效地吸收烘焙过程中产生的有害气体。这种安全环保的功能是我选择这款产品的原因之一。

B：你作出了明智的决定。无烟烘焙对我们的健康非常重要。

A：现在烘焙完了。我们把这个密封手柄移到左边，把银皮收集罐拿出来。

B：我喜欢它的味道。你注意到颜色了吗？看起来像是中度浅棕色。对吗？

A：对。烘焙完的豆在这个大箱子里，而豆壳在这个小箱子里。这种分离设计确实节省了时间和精力。

B：完全正确。与那个大而复杂的烘焙机相比，这个小型自动烘焙机更适合我。

A：约翰，我想买些烘焙好的咖啡豆。你有什么建议吗？

B：有。烘焙好的咖啡豆的主要敌人是氧气、水分、光、热和强烈气味，所以尽量选择有盖子或单向气阀的容器保护的包装。例如，带有单向气阀的不透明密封袋就是个不错的包装。

A：有些咖啡豆是装在牛皮纸袋或真空密封袋里的。

B：在我看来，牛皮纸袋只能为咖啡提供最低限度的保护。真空密封的方法去除了咖啡中过多的香气挥发物，这对其感官特性产生了负面影响。我不会选择这些包装的。

A：我明白了。还有其他需要我注意的吗？

B：记得看烘焙日期。一些专家认为，烘焙两周以上的咖啡豆已不新鲜了。对此我不确定。不过尽可能买新鲜的东西总是对的。

A：好的，我明白了。

B：你喜欢什么口味的？

A：我不知道。

B：在这种情况下，试试尽可能多地了解产品的信息。你可以从包装和咖啡师那里得到有关信息。挑几种买一点来比较，试一试，直到你找到你想要的质量和味道。

A：包装上会有什么信息？

B：理想情况下，你可以找到生产商和（或）农场主的名称、风味、咖啡处理方式、海拔高度、烘焙程度、烘焙和包装日期和出处，这些信息将告诉你咖啡的种类和（或）品种，无论是拼配咖啡还是单一产区咖啡等。

A：太复杂了。我可以直接选贵的咖啡豆吗？

B：恐怕这不是一个比较好的选择。找咖啡师指导一下，看看有什么建议吧。他们对咖啡豆的选择和设备的使用都非常专业。

A：好啊。你愿意和我一起去吗？

B：不好意思，我忙得不可开交。明天去怎样？

A：没问题。

任务模块4　知识拓展

了解更多关于烘焙咖啡生豆的信息

在烘焙咖啡生豆的过程中，可以听到烘焙机发出的爆裂声。当温度达到约196 ℃（385 ℉）时，蒸汽的压力最终会破坏豆子的细胞结构，并第一次发出开裂声。咖啡豆的尺寸增大，外观颜色变得与肉桂一样均匀，闻起来开始像咖啡。草味已被除去，但烤谷物味明显。香气闻起来很好，酸度很高，但甜度还没有开发出来。豆子进入一个非常轻的烘烤阶段，通常称为轻中度烘焙（肉桂烘焙）阶段。它通常用于美式咖啡中。

当温度接近224 ℃（435 ℉）时，豆子会因为气体压力产生第二次爆裂。豆子完全变成脆性的，颜色变为中度深棕色。豆子表面会出现微小的油滴或模糊的油斑。烘焙的特性变得突出，苦大于酸。中度深棕色水平通常称为深城市烘焙阶段。

　　烘焙生豆的不同阶段产生了不同的咖啡豆特性。除上述两个阶段外，其他6个阶段咖啡豆的特性如下。

序号	常见烘焙阶段	特征	烘焙程度	风味
1	轻度烘焙	烘焙程度最轻，开始变成黄色。由于味道和香气不足，这一阶段的豆子很难饮用，通常用于检验。	轻度烘焙	豆子较轻，酸度较高，几乎没有苦味，没有明显的烘烤味道。主要是花香和水果味。咖啡的来源特性占主导地位。
2	肉桂烘焙（轻中度烘焙）	豆子进入一个烘焙程度较低的阶段。第一次爆裂声从烘焙炉发出。豆子的颜色变得和肉桂一样均匀。美拉德反应没有充分发挥作用，酸度很高。		
3	中度烘焙	浅棕色，但表面有斑点，酸度明亮，香气适中，有少许苦味。	中度烘焙	随着进一步焦糖化，其他成分相对较少。酸度和甜度得到了强调，几乎没有烘烤风味。特别适合制作单品咖啡。
4	深度烘焙	中度棕色，酸苦相对均衡，香气和风味良好，但仍保留了原汁原味。		
5	城市烘焙	棕色，苦味和酸度良好平衡，起源特性和烘焙特性和谐共存。		
6	深城市烘焙	第二次爆裂开始，偏深棕色，表面带细小的油斑，烘焙特征很明显，苦味比酸度强。	深度烘焙	有光泽、轻薄，大部分糖都烧掉了，剩下一点酸度，烘烤的香气和味道变得明显。
7	法式烘焙	深棕色，表面因油脂而发亮，咖啡豆的酸度和几乎所有固有的香气或味道都基本消失，苦味非常强烈，烘焙特性很突出。		
8	意式烘焙	表面基本是黑色的，豆身轻，表面有光泽，咖啡豆的原味消失了，而焦糊味变得非常明显，几乎只有烘烤、灰味和苦味，没有酸度。		

注意：根据研究，咖啡因在200 ℃（392 ℉）以下保持稳定，在285 ℃（545 ℉）左右完全分解。鉴于烘焙温度通常长时间不超过200 ℃（392 ℉），很少达到285 ℃（545 ℉），咖啡豆中的咖啡因含量在烘焙过程中可能不会有太大变化。

任务模块5　创意任务

许多特色咖啡馆会储存各种烘焙咖啡豆，以便为顾客提供单一产区咖啡或拼配咖啡做好准备。两人一组，尝试向其他人介绍不同的烘焙程度以及常见的烘焙名称，以帮助他们更多地了解烘焙技术的特点。如有必要，我们鼓励你们从图书馆或互联网搜集更多信息。

任务模块6　任务执行能力考核

序号	考核细分项目	标准分数（分）	得分（分）
1	专业用语	10	
2	任务资料的基础训练	20	
3	场景任务1专项训练	20	
4	场景任务2专项训练	20	
5	知识拓展	10	
6	创意任务	20	

任务模块7　目标达成考核

评分考核项目	标准分数（分）	个人自评	小组互评	教师评分
理论知识	30			
技能训练	40			
职业素养	30			
总分（分）	100			
综合总分				
说明	综合总分=个人自评（占总分的20%）+小组互评（占总分的20%）+教师评分（占总分的60%）			

任务3 咖啡研磨

 ## 第1部分 任务介绍

咖啡研磨

1. 教学目标

知识目标	1. 了解咖啡的研磨方法。 2. 了解手动磨豆机和自动磨豆机的区别，学习磨豆机的基本操作流程。 3. 了解不同粗细度咖啡粉的区别及其适用的冲煮方法。
技能目标	1. 理论联系实际，通过咖啡豆研磨操作训练，熟悉基本操作流程，并研究绘制出不同研磨方法的操作流程图。 2. 学会辨别咖啡粉的粗细度，构思设计咖啡粉与冲煮方法的匹配图。 3. 在学习英语的过程中，感受英语思维方式，提升思维的逻辑性、思辨性和创新性。 4. 能够运用英语比较准确地理解和表达信息、观点、情感，能有效地进行跨文化交际与沟通。
素质目标	1. 培养与人协作的精神和团队合作的意识。 2. 拓宽国际视野，尊重世界多元文化，树立人类命运共同体意识。 3. 通过辨别与对比能力训练获得认同感，提高自信心，促进正确英语交际观的养成。 4. 做训结合，从不同角度加强训练，构建设计与创新思维。 5. 领会咖啡豆研磨的关键点以及咖啡粉粗细度对咖啡品质的影响，培养精益求精、注重品质的工匠精神。

2. 教学重点

（1）了解咖啡研磨的基本操作流程。

（2）对比咖啡研磨方法，领会研磨度与冲煮方法的关系。

3. 教学难点

能用英语进行跨文化交际与沟通。

第2部分　任务实施

任务模块1　前置任务

步骤1　小组合作

两人一组，讨论下列方框中的专业用语并加以熟悉。如果需要，你可以使用智能手机在线查找。

> coffee grinder　manual grinder　electric grinder　grinding degree
> uniformity of coarseness

步骤2　小组讨论

两人一组，讨论以下问题并交流观点。

1. 你见过或使用过哪些类型的咖啡研磨机？
2. 手动咖啡研磨机和电动研磨机有什么区别？
3. 选择咖啡研磨机时，你会考虑哪些因素？

任务模块2　阅读与训练

咖啡研磨

如果你想提高咖啡的质量并获得较好的质地，有一种非常简单的方法可以让你做到，那就是用一个好的磨豆机研磨新鲜的咖啡豆。市场上有两种类型的磨豆机。一种是手动磨豆机或手动研磨机。另一种是电动磨豆机。与电动磨豆机相比，手动磨豆机体积小，重量轻，便于携带，易于清洁。随着技术的不断创新，特别是可调整的刀盘的应用，手动磨豆机可以更加方便地调整粗糙度的均匀性。手动磨豆机可以使咖啡粒度更均匀。研磨过程中不会过热，有助于保持咖啡的原味和香气。虽然它通常只有一两个储粉罐，但在家里、办公室、旅行或露营时，它受到人们的青睐而被广泛应用。你可以在煮水时慢慢磨咖啡，然后冲泡一杯浓而滑的咖啡，度过一个轻松愉快的早晨。许多人真的很喜欢这个过程。手工磨豆机开始拥有自己的市场份额。

然而，尽管电动磨豆机比手动磨豆机更昂贵，但它们仍然占据着主要市场，因为它们可以轻松、快速地研磨更多的咖啡。电机产生的摩擦和废热可能会轻微加热咖啡粉，从而影响咖啡的质量。但技术的发展使这种影响越来越小。在当今市场上，你可以为酒店、车库、家庭、咖啡馆或办公室购买不同容量的电动磨豆机。无论你想要什么类型的，你总是可以买一个你喜欢的。

那么如何选择咖啡磨豆机呢？需要考虑很多因素，包括价格、研磨量、大小、研磨度、研磨粗细均匀度、材料、产生的噪声、刀盘或刀片等。研磨度和研磨粗细均匀度被认为是提高咖啡质量的两个最重要的因素。

任务模块3　场景交际任务

场景1

A：凯拉，如何使用咖啡磨豆机？

B：你想用哪一个？

A：这两个有什么区别？

B：它们都是电动磨豆机，这款是用于滴漏咖啡的，那款是用于意式咖啡（浓缩咖啡）的。

A：你能具体说说吗？

B：可以。这款用于滴漏咖啡的磨豆机可以调节研磨度。它使用快捷方便，但不能研磨精细或超精细的咖啡粉。

A：那这个呢？看起来是新款。

B：这是用于意式咖啡的磨豆机。我们去年买的，因为我们经常需要研磨大量的咖啡粉来做咖啡。

A：告诉我这个怎么操作。

B：好。这是储存豆子的豆仓。这是控制研磨粗细度的刻度调节器。这是数字定时功能。屏幕上的数字是研磨时间。你可以使用这个小按钮进行时间调整。橙色按钮控制整个机器的开和关。黑色按钮是"开始研磨"。现在让我教你怎么操作。把豆子递给我。

A：给你。

B：第一步是将20克咖啡豆倒入豆仓。下一步按橙色按钮启动机器。这个屏幕亮了。然后，如果有必要，可以转动调节器来调整研磨的粗细度。现在它指向4。你可以帮我把它调到8吗？

A：让我试试。是这样吗？

B：是的，非常好。接着调整时间。例如，我们用这个按钮把时间设成6秒钟。现在按下黑色按钮开始研磨。

A：哇，它磨起来了。

B：时间到了，机器就会自动停止。

A：真有趣。我可以自己磨一些咖啡豆吗？

B：可以，你试试。

场景2

A：凯拉，为什么我们需要调整刻度和时间？

B：让我们做一个实验。我们将同样数量的豆子分3次研磨，用不同刻度但相同时间。然后我们对咖啡粉进行比较。

A：好的。好想快点看到结果。

B：现在第一次研磨，用刻度4在6秒钟内研磨。磨完将咖啡粉倒入A杯。

A：好的。我把杯子标记为A。

B：现在是用刻度8在6秒钟内研磨。

A：我把粉倒进B杯。

B：很好。现在用刻度12在6秒钟内研磨。

A：最后这些粉倒入C杯。

B：他们之间有什么不同？

A：A杯中的咖啡粉颗粒最小，B杯中的颗粒中等，C杯中的颗粒最大。

B：是的。我们将A杯的咖啡粉称为细粉，B杯的咖啡粉称为中度粉，C杯的咖啡粉称为粗粉。如果相同数量的豆子以相同的研磨刻度但在不同的时间内研磨，结果会跟这个一样。你用的时间越短，咖啡粉就越粗糙。反之亦然。

A：我明白了。

B：完成研磨后，需要清洁磨豆机。你愿意和我一起清理机器吗？

A：我很乐意。

任务模块4　知识拓展

了解更多关于咖啡研磨的信息

无论是机器研磨还是手工研磨，咖啡豆都可以用多种方式研磨。有磨盘的机器称为刀盘式磨豆机，它使用旋转的磨盘元件来粉碎咖啡豆。带刀片的机器被称为桨叶式磨豆机，它使用高速转动的刀片来切碎咖啡豆。对于大多数咖啡冲煮方法来说，刀盘式磨豆机被认为是更优秀、更受欢迎的选择，因为它可以产生颗粒大小均匀的咖啡粉，而刀片通常会将咖啡豆切成大小不一致的粉末。咖啡粉也可以用研钵、杵或锤子来粉碎而成。过去，牛仔们会用锤子把咖啡豆敲碎，然后用锅在篝火上煮咖啡。

你知道哪种研磨方法更适合咖啡冲泡方法吗？按咖啡粉的粗细度来分，可以有4种研磨方式。它们是超细粉研磨（也称土耳其式研磨）、细粉研磨、中度粉研磨和粗粉研磨。一些鉴赏家认为，超细粉研磨可以产生最精细的咖啡粉。最好使用土耳其伊布里克咖啡壶萃取粉末的最大风味。细粉研磨会磨出细的咖啡粉，非常适合浓缩咖啡机进行适当的萃取。中度粉研磨为你提供中等颗粒度的咖啡粉，可以用于多种咖啡冲煮方法，如手冲咖啡、火炉式咖啡、法压式咖啡、虹吸式咖啡等。粗粉研磨为你提供最粗的咖啡粉。这对于使用法兰绒滤杯或滴漏咖啡机等的冲泡方式来说更合适。使用这些器具时，水可以穿透粗粉的细胞结构。这种方法可以溶解令人愉快的可溶性物质，避免过多的苦味。

任务模块5　创意任务

定期清洁咖啡磨豆机是有效保护设备的关键。如何清洁？试着和伙伴一起清洁磨豆机，并写出流程。在最后确定之前，请与伙伴讨论你写的流程。我们鼓励你根据需要从图书馆或互联网获取更多信息。

任务模块6　任务执行能力考核

序号	考核细分项目	标准分数（分）	得分（分）
1	专业用语	10	
2	任务资料的基础训练	20	
3	场景任务1专项训练	20	
4	场景任务2专项训练	20	
5	知识拓展	10	
6	创意任务	20	

任务模块7　目标达成考核

评分考核项目	标准分数（分）	个人自评	小组互评	教师评分
理论知识	30			
技能训练	40			
职业素养	30			
总分（分）	100			
综合总分				
说明	综合总分=个人自评（占总分的20%）+小组互评（占总分的20%）+教师评分（占总分的60%）			

Project Three

Coffee Preparation

Task 1   Cupping

Part One   Task Introduction

1. Teaching Objectives

Knowledge Objectives	1. Understand the development of coffee cupping. 2. Grasp the type and function of coffee cupping. 3. Through learning, understand the professional standards of cupping, and master the necessary professional knowledge in flavor wheel and sensory vocabulary.
Skill Objectives	1. Be able to distinguish the instruments required for cup measurement, and show their use by operating the cup measurement process. 2. In the process of learning English, feel the way of thinking in English, and improve the logic, speculation and innovation of thinking. 3. Be able to use English to understand and express information, views and feelings more accurately, and conduct effective cross-cultural communication.
Competency Objectives	1. Cultivate the sense of collaboration and teamwork. 2. Learn the detailed and professional specifications of flavor wheel and sensory vocabulary, and cultivate rigorous, meticulous and standardized professionalism. 3. Learn from the cupping tester or appraiser's practical and realistic professional attitude, as well as the craftsman's spirit of hard work and excellence. 4. Learn to choose learning resources and use appropriate English learning strategies for autonomous learning.

2. Teaching Focuses

(1) The coffee cupping tools and uses.

(2) Standardized operation process for coffee cupping.

3. Teaching Difficulty

Effective intercultural communication in English.

 咖 啡 (双语)

 Part Two   Task Implementation

Task Module 1   Pre-tasks

Step 1   Group Cooperation

Work in pairs. Discuss the professional terms in the box and get familiar with them. You can look them up online with your smartphones if necessary.

> cupping   tasting   cupper   G-grader   SCA   evaluate   defective bean

Step 2   Panel Discussion

Work in pairs to discuss the following issues and exchange views.

1. How do people evaluate the quality of coffee?

2. What kinds of standards will be used when people evaluate coffees?

Task Module 2   Reading and Training

Step 1   Reading

Coffee Cupping

Coffee cupping, or coffee tasting, is a method of evaluating coffees. According to the whole industrial chain of coffee from farm to cup, there are cuppings at place of origin, cuppings for bulk sales of green beans and cupping for roasted beans.

Cupping at place of origin is the most elementary cupping which is often used by bean seekers to help them determine the prices of these coffee beans. It will be evaluated according to different regions. The most important aspect of cupping in this stage is to find out the local specific flavor characteristics of these beans and check out whether these beans are polluted or not.

After cupping, the beans will be graded as per different standards, like Ethiopian standard for grading by the proportion of defective beans (G1-G5), Kenya standard for grading by particle size (E, AA, AB, C, PB, TT, T, MH/ML), etc. After grading, coffee will be sold at different prices on the basis of different grades. For this stage, the results are mainly related to defect rate, size, growth altitude, hardness and the like, but not much related to coffee flavor. Cupping for roasted beans is the practice of evaluating characteristics by tasting and measuring the coffee which will assist in the decision of the roast profiles, quality control and buying decision.

It's believed that cupping is a practice used for more than one hundred years. It could be originated back to the late 1800s when merchants used the method as a way of running consistency checks during the buying process. At the end of the 20th century, it became widely used after the Cup of Excellence competition began using cupping to judge and evaluate coffee samples. Later, a cupping protocol created by the Specialty Coffee Association (SCA) of America became a popular guideline which remains standard for today's coffee community.

Cupping is a professional practice carried out by G-graders. But the relatively simple technique can also be done informally by anyone. Some experts think that as cupping is based on a sensory evaluation, it is subjective. Whether the result is accurate will be influenced by the ability of tasters. So they suggest that it will be ideal to combine both human cupping (strictly following the SCA protocol by professionals) and technology (like near-infrared spectroscopy and AI) to produce reliable cupping results.

Word and Phrase Bank

New Words

elementary /ˌelɪ'mentri/ *adj.*　简单的，基本的

determine /dɪ'tɜ:mɪn/ *v.*　决定

pollute /pə'lu:t/ *v.*　污染

proportion /prə'pɔ:ʃn/ *n.*　部分，比例

defective /dɪ'fektɪv/ *adj.*　有问题的

altitude /'æltɪtju:d/ *n.*　海拔高度

hardness /hɑ:dnəs/ *n.*　硬度

merchant /'mɜ:tʃənt/ *n.*　商人

consistency /kən'sɪstənsi/ *n.*　一致性

protocol /'prəʊtəkɒl/ *n.*　协议；规章制度

sensory /'sensəri/ *adj.*　感觉的

subjective /səb'dʒektɪv/ *adj.*　主观的

accurate /'ækjərət/ *adj.*　正确的

Phrases & Expressions

coffee cupping　杯测

industrial chain　产业链

bulk sales　整批出售

check out　检验

Cup of Excellence competition　"超凡杯"比赛

the like　类似的东西

carry out　执行

near-infrared spectroscopy　近红外光谱学

Step 2　After-reading Training

Ⅰ　Choose the best answer to complete each statement.

1. Coffee cupping, or coffee tasting, is a method of _____.
    A. weighting coffees   B. tasting coffees
    C. evaluating coffees   D. seeking coffees

2. Cupping at _____ is the most elementary cupping which is often used by the local exchange officials or bean seekers.
    A. place of origin   B. bulk sales   C. roasted beans   D. flavor

3. Cupping for bulk sales of green beans is used for the beans transaction in _____.
    A. barrel   B. bag   C. jar   D. bulk

4. Ethiopian standard for grading by the proportion of _____.
    A. particle size   B. color   C. defective beans   D. weight

5. Coffee will be sold at different prices related to defect rate, size, _____, _____ and the like.
    A. hardness, color   B. odor, growth altitude
    C. color, bitterness   D. growth altitude, hardness

Ⅱ　Read the statements and tick the correct boxes.

Statements	Right	Wrong	Not mentioned
Cupping for roasted beans is the practice of evaluating characteristics by tasting and measuring the green beans.			
It's believed that cupping is a practice used for more than five hundred years.			
Cupping could be originated back to the late 1800s when the merchants used the method as a way of running consistency checks during the buying process.			
Cupping in the Cup of Excellence competition can be done by some simple techniques.			
Some experts suggest that it will be ideal to combine both human cupping and technology to produce reliable cupping results.			

Answers to the Tasks

Task Module 3　Communication Tasks

Situation 1

Step 1　Role-playing

A: Natalie, what are you doing?

B: I'm preparing for the coffee cupping.

A: Why?

B: We would like to know the roasting effect of the beans we roasted yesterday.

A: What are all these things for?

B: This is water with temperature normally between 92.2–94.4 ℃.

A: What is this?

B: It's a grinder. We'll use the medium-fine grind size of coffee sample for cupping so as to maximize the extraction.

A: Then what are these cups used for?

B: We call these cupping cups. Although SCA recommends the standard cupping cups with maximum capacity of 220 mL, I think any cup with 110–300 mL capacity will work. Two things should be noticed. One is the ratio of coffee to water. For our cups, we use 8.25 g coffee beans with 150 mL of water. For the cups of 220 mL, 12 g coffee beans will be perfect. The other thing is to use the cups or bowls with a lid. When you cover it on the cup or bowl, it could help to keep the aromas of ground coffee.

A: How about others?

B: This is a cupping spoon. It will be used to stir the crust.

A: What is "crust"?

B: After we add hot water to the coffee grounds, soon there will be a layer of grounds floating on the surface. We call it "crust".

A: I see.

B: And we also use the spoon to skim off the remaining foam from the surface.

A: What else?

B: The large glass filled with water is for rinsing the spoon. And there is a digital scale, a timer and a notepad.

A: Why do you need the notepad?

B: Look, here are some cupping forms. The tasters are required to take detailed notes about aspects of the coffee's quality, mainly including Fragrance/Aroma, Flavor, Acidity, Body, Uniformity, Clean Cup, Balance, Sweetness, Defects, Overall, etc.

A: Wow, it's really complicated for me.

 **Word and Phrase Bank**

New Words

extraction /ɪkˈstrækʃn/ *n.* 提炼；萃取

capacity /kəˈpæsəti/ *n.* 容积

aroma /əˈrəʊmə/ *n.* 芳香

crust /krʌst/ *n.* 硬外壳，咖啡颗粒

notepad /pæd/ *n.* 记事本

Phrases & Expressions

medium-fine 中细的

skim off 从……挑选出

Step 2　Consolidation Training

I　Match the words on the left with their meanings on the right.

1. sample	A. a particular part or feature of a situation, an idea, a problem, etc.
2. grinder	B. left over after a part has been destroyed, taken, used, or lost
3. extraction	C. a representative part or a single item from a larger whole or group especially when presented for inspection or shown as evidence of quality
4. recommend	D. reduced to small pieces or a powder by a grinding process
5. lid	E. the state of having a sour taste or of containing acid
6. ground	F. a machine or device for grinding
7. remaining	G. to suggest that a particular action should be done
8. aspect	H. a taste (usually an unpleasant one) that stays in your mouth after you have eaten or drunk something
9. acidity	I. something extracted
10. aftertaste	J. a cover on a container that can be lifted up or removed

1. _____　2. _____　3. _____　4. _____　5. _____　6. _____　7. _____

8. _____　9. _____　10. _____

II　Decide whether the following statements are true (T) or false (F).

(　　) 1. The fine grind size of coffee sample is used for cupping so as to maximize the extraction.

(   ) 2. SCA recommends the standard cupping bowls with maximum capacity of
          220 mL.

(   ) 3. After we add cold water to the coffee grounds, soon there will be a layer of
          grounds floating on the surface.

(   ) 4. We use the spoon to skim off the remaining grounds or foam on the surface.

(   ) 5. The tasters are only required to take notes about the acidity and flavor of the
          coffee's taste.

Situation 2

Answers to the Tasks

Step 1   Role-playing

A: Natalie, how do you perform cupping?

B: Let me show you just one cup. The first step is to grind the coffee samples. Weigh 8.25 g coffee beans. Then pour the ground coffee powder into the cup.

A: How about the next step?

B: Now we sniff the dry ground coffee and evaluate them. We bend down and sniff. Remember: don't touch the cups with your hands.

A: Why?

B: The smell like cigarette, hand cream or liquid soap on your hands will be left on the cups. It will influence our identification and evaluation.

A: Does that mean anything with smells like perfume, make-up or cigarette should be avoided?

B: Absolutely correct. Then we need to make records on the cupping form. After that we'll start to pour 94 ℃ water into the cup. To ensure that the ground coffee is completely soaked without caking, fill the cup with a large flow of water. The ratio of coffee to water is generally 1:18.18.

A: Your operation is very skilled.

B: Thank you. When we start to pour water, start the timer at the same time. The best way is to count down with the timer for 4 minutes. When the time comes, it will beep to remind you.

A: OK. What should I do now?

B: Let's just wait. 1 minute later, we'll sniff the aroma of the wet coffee. Be careful. They're very hot. Then write down all your feelings on the cupping form.

A: Got it. Wow, this one smells very well.

B: The ground crust will be formed in 4 minutes. When the timer beeps, we'll break the crust together.

A: How to break it?

B: Like this. Puncture the crust with 3 times. Then inhale the aromatics. Good. The next step is to skim off the foam and drag with two cupping spoons simultaneously. Remember you should always rinse the spoon and click the tip of the spoon on the paper before you move to the

next cup. Cross-contamination is not allowed.

A: OK. Rinse, click and the next cup. I got it. When can I taste the coffee?

B: After 8–10 minutes of infusion, when the temperature cools down to about 50–60 ℃, evaluation should begin. Now scoop the liquid into your mouth and quickly slurp so as to cover as much area of your mouth as possible.

A: What should I focus on especially?

B: The coffee's body, acidity, sweetness, and flavor. You could repeat the steps for several times if necessary. The final step is to take notes.

A: It doesn't seem difficult, but it's not so easy to do it accurately.

B: Yes. In order to be accurate, the professional cuppers or tasters need not only keen senses, but also a lot of practice.

Word and Phrase Bank

New Words

sniff /snɪf/ *v.* 嗅，闻

identification /aɪˌdentɪfɪ'keɪʃn/ *n.* 辨认，识别

puncture /'pʌŋktʃə(r)/ *v.* 刺破，戳破

inhale /ɪn'heɪl/ *vt.* 吸入，吸气

simultaneously /ˌsɪm(ə)l'teɪniəsli/ *adv.* 同时地

slurp /slɜːp/ *v.* 啜饮

accurate /'ækjərət/ *adj.* 准确的，精确的

Phrases & Expressions

bend down 蹲下，弯腰

liquid soap 洗手液

count down 倒计时

Step 2　Consolidation Training

Ⅰ　Complete the following sentences with the words in the box. Change the form if necessary.

> influence　identification　operation　contamination　accurately

1. What skills are needed to _____ this machine?

2. This chemical factory might _____ the whole area.

3. Those friends are a bad _____ on her.

4. _____ measurement is very important in science.

5. We must _____ the problem areas.

Ⅱ   There are ten English words and terms numbered 1 through 10 in the left column. You should match them with their Chinese equivalents marked A through J in the right column and write down the corresponding letter in the right blank.

1. grounded coffee	A. 避免
2. hand cream	B. 结块
3. make-up	C. 咖啡粉
4. avoid	D. 同时地
5. caking	E. 拖动
6. fill with	F. 护手霜
7. cross-contamination	G. 化妆品
8. drag	H. 浸泡
9. simultaneously	I. 装满
10. infusion	J. 交叉污染

1. _____   2. _____   3. _____   4. _____   5. _____   6. _____
7. _____   8. _____   9. _____   10. _____

Answers to the Tasks

Task Module 4   Extensive Reading
Learn More about Cupping of Coffee

Coffee is considered to be one of the most chemically complex beverages in the market. Coffee's aroma, texture and flavor can hardly be compared with other foods. There are many elements that will have an effect on the quality of coffee, like the seed's genes, the place and the way it grows, processing, storage, transport, roasting and brewing, etc. So, in order to find out the coffee that we want, we need to try to smell, drink, taste, feel and evaluate coffee. The cupper or taster appeared as the times required.

In order to create a standard for barista competitions and a reference for others to measure a coffee, the Coffee Taster's Flavor Wheel was created in the late 1990s for the SCA by Ted Lingle. With the purpose of creating a universal language of coffee's sensory qualities and a tool for measuring them, World Coffee Research organized a group of 10 sensory scientists from Kansas State University using the Sensory Analysis Center at Kansas State University, one of the world's premier sensory science centers, to do the repetitive researches of coffee

taste. The World Coffee Research Sensory Lexicon which grew to 110 attributes was developed and published in 2016. As new varieties of coffee are continually being cultivated, the above documents are called as living documents. Thanks to these documents, coffee cupping and tasting could be more professional, more easily descriptive and reliable.

Task Module 5　Creative Task

Work in a group. Try to find out the documents of Coffee Taster's Flavor Wheel or the World Coffee Research Sensory Lexicon. Learn how to use them and share your study with others. Maybe there are quite a lot of new words and expressions. Don't worry. You are encouraged to collect more information from the library or through the internet if necessary.

Task 2   Brewing

Part One   Task Introduction

1. Teaching Objectives

Knowledge Objectives	1. Understand the different methods of brewing coffee and the factors that should be considered when brewing coffee. 2. Master the characteristics and basic principles of common coffee brewing equipments. 3. Master the characteristics of espresso and the ingredients of common espresso drinks. 4. Identify the main components and functions of the semi-automatic espresso machine.
Skill Objectives	1. Combine theory with practice, and be able to correctly operate the semi-automatic espresso machine. 2. In the process of learning English, feel the way of thinking in English, and improve the logic, speculation and innovation of thinking. 3. Be able to use English to understand and express information, views and feelings more accurately, and conduct effective cross-cultural communication.
Competency Objectives	1. Cultivate the sense of collaboration and teamwork. 2. Through the operation training of the semi-automatic espresso machine, cultivate the craftsmanship spirit of a barista to keep safe operation, strict specification, concentration and carefulness in mind during coffee brewing. 3. Learn the ingredients and formulas of different coffee drinks, and cultivate the innovative consciousness of active innovation, pursuit of quality and breakthrough. 4. Build confidence and set up a correct view of English learning.

2. Teaching Focuses

(1) Features and differences of common coffee brewing equipment.

(2) Standard operation of semi-automatic espresso machine.

(3) The characteristics and innovation of coffee drinks.

3. Teaching Difficulty

Effective intercultural communication in English.

Part Two   Task Implementation

Task Module 1   Pre-tasks

Step 1   Group Cooperation

Work in pairs. Discuss the professional terms in the box and get familiar with them. You can look them up online with your smartphones if necessary.

> coffeemaker beverage decoction steep pressurization extraction filter

Step 2   Panel Discussion

Work in pairs to discuss the following issues and exchange views.

1. Do you know how many types of coffeemakers there are? And what are the differences among them?

2. Do you know how to brew a cup of coffee? Have you ever made it by yourself?

Task Module 2   Reading and Training

Step 1   Reading

Brewing

Coffee brewing is the process of turning coffee grounds into a beverage. It is suggested that coffee should be brewed before drinking. People from different areas may brew coffee in different ways. For example, Turkish coffee also known as Arabian coffee is considered to be the ancestor of today's coffee. Turkish coffee is very finely ground coffee brewed by boiling. Turkish coffee is popular in countries such as Turkey, Greece and Balkans. The Indian filter coffee is particularly common in southern India.

Boiling, or decoction, is the method to cook coffee grounds with cold water. For the Turkish coffee, only very finely ground coffee will be brewed. It could be finished by boiling over for 3 times and served in dark brown with dense foam on the surface and the coffee grounds at the bottom. Same by boiling method, the "Cowboy coffee" in the early days was made by heating coarse grounds with cold water in a pot or a can in the wild and drank with sugar or spices to cover its bitter and negative taste.

92

Steeping is considered as a relatively traditional, natural and simple way which can be further divided into artificial pressure filtration, vacuum brewing, drip filter and steam pressurization of Moka pots. It brews coffee grounds with hot water normally at about 94 ℃ (201 °F) under about a pressure of 1 bar. Pressurizing refers in particular to Italian espresso machine which is to make coffee in about 25 seconds by forcing hot water at 92–96 ℃ (198–205 °F) under a pressure of about 9 bars through finely ground coffee. The two methods both require to mix the ground with hot water and soak it in a limited time for the excellent flavor emerging but with as little bitterness as possible. And the coffee liquid will be served without the spent grounds. It is believed that the relatively ideal temperatures of coffee brewing and serving are: between 92–96 ℃ (198–205 °F) for grounds extraction; from 85 ℃ to 93 ℃ (185 °F to 199 °F) for soaking grounds; from 68 ℃ to 79 ℃ (154 °F to 174 °F) for coffee serving.

When brewing coffee, the following factors are usually considered: quantity of ground coffee, the fineness of grounds, uniformity of grounds coarseness, water temperature the brew ratio (the ratio of coffee grounds to water), the way of grounds extraction (how the water is used to extract the flavor), the desired additional flavorings (like sugar, milk and spices), the separation technique (separating the liquid and spent grounds), etc.

 Word and Phrase Bank

New Words

beverage /'bevərɪdʒ/ *n.* 饮料

ancestor /'ænsestə(r)/ *n.* 雏形

Balkans /'bɔːlkənz/ *n.* 巴尔干半岛地区

decoction /dɪ'kɒkʃən/ *n.* 煎煮

filtration /fɪl'treɪʃn/ *n.* 过滤

method /'meθəd/ *n.* 方法

extraction /ɪk'strækʃn/ *n.* 抽出

uniformity /ˌjuːnɪ'fɔːməti/ *n.* 一致（性）

separation /ˌsepə'reɪʃn/ *n.* 分离

Phrases & Expressions

ground coffee 磨碎的咖啡

boiling over 沸腾

negative taste 不好的味道

drip filter 滴滤

steam pressurization 蒸汽定压

Moka pot 摩卡壶

additional flavoring　附加调味品

separation technique　分离技术

Step 2　After-reading Training

Ⅰ　Answer the questions.

1. What is coffee brewing?

_____

2. What is the ancestor of today's coffee?

_____

3. How many operations of steeping are there?

_____

4. What are the relatively ideal temperatures of coffee brewing for grounds extraction?

_____

5. What factors are usually considered when brewing coffee?

_____

Ⅱ　Read the statements and tick the correct boxes.

Statements	Right	Wrong	Not mentioned
Coffee brewing is the process of turning coffee grounds into a beverage.			
For the Turkish coffee, only very finely ground coffee will be brewed.			
"Cowboy coffee" in the early days was made by heating coarse grounds with hot water in a pot or a can in the wild and drank with milk to cover its bitter and thick negative taste.			
Steeping is considered as a relatively traditional, natural and simple way which can be further divided into artificial pressure filtration, vacuum brewing, drip filter and steam pressurization of Moka pots.			
It is believed that the relatively ideal temperatures of coffee brewing and serving are: between 85 ℃ to 93 ℃ (185 °F to 199 °F) for grounds extraction.			

Answers to the Tasks

Task Module 3   Communication Tasks

Situation 1

Step 1   Role-playing

A: Good morning, Alex. Wow, what are these parcels?

B: They are different coffeemakers. Can you help me to unpack them?

A: No problem. What is the biggest parcel?

B: It is a commercial semi-automatic coffee machine.

A: What are those then?

B: They are Chemex, French press, AeroPress, siphon pot, and...

A: Wait. I'm totally confused.

B: Let me tell you another way. Generally speaking, there are two main types of coffeemakers. The first type is just like this one. It is Italian espresso machine which could be semi-automatic and automatic.

A: What's the difference between them?

B: With the automatic machine, you don't need to manually grind and tamp the coffee. It will automatically extract the espresso shot as well. But for semi-automatic one, these first two steps have to be done by yourself. And you need an extra coffee grinder to grind the grounds, too.

A: I got it. How about the other type?

B: We call it coffeemakers. The coffeemakers are used to make coffee by boiling or steeping extraction. People usually subdivide steeping extraction into artificial pressure filtration, vacuum brewing, drip filter and steam pressurization.

A: Could you give me some examples?

B: Yes. This is French press and AeroPress which belong to artificial pressure filtration. You need to brew the coffee with artificial pressure to separate the coffee grounds from the liquid. It means you need to press it by hand like this.

A: OK.

B: Now is the siphon (or syphon) coffeemaker. It makes use of the pressure difference caused by steam cooling down to filter coffee. It is a vacuum coffeemaker.

A: How about this one with a cup?

B: It is Hario V60. Hario V60 and this one called Chemex are for pour-over. They are using the natural gravity to drip and filter coffee. So we put it into the type of drip filter.

A: Then this silver one.

B: It's a Moka pot which brews coffee by passing boiling water pressurised by steam through ground coffee. So people classify it as a steam pressurized coffee pot.

A: I basically understand. Could you show me how to use them?

B: My pleasure. But firstly, we need to clean them and find rooms for each one.

 **Word and Phrase Bank**

New Words

parcel /'pɑːsl/ *n.* 包裹

unpack /ˌʌn'pæk/ *v.* 打开取出

commercial /kə'mɜːʃl/ *adj.* 商业的

grinder /'graɪndə(r)/ *n.* 研磨器

filtration /fɪl'treɪʃn/ *n.* 过滤

gravity /'grævəti/ *n.* 重力

silver /'sɪlvə(r)/ *adj.* 银色的

Phrases & Expressions

drip filter   滴槽滤网，滴滤

as well   也，还

vacuum brewing   真空冲泡

artificial pressure   人工增压

be subdivided into   细分为

Step 2   Consolidation Training

I   There are ten English words and terms numbered 1 through 10 in the left column. You should match them with their Chinese equivalents marked A through J in the right column and write down the corresponding letter in the right blank.

1. grinder	A. 过滤
2. gravity	B. 商业的
3. parcel	C. 压力
4. silver	D. 包裹
5. filtration	E. 额外的
6. commercial	F. 打开取出
7. pressure	G. 研磨器
8. extra	H. 人工的
9. unpack	I. 重力
10. artificial	J. 银色的

1. _____   2. _____   3. _____   4. _____   5. _____   6. _____   7. _____

8. _____   9. _____   10. _____

Ⅱ   Decide whether the following statements are true (T) or false (F).

(   ) 1. With the automatic machine, you need to manually tamp and grind the coffee.

(   ) 2. French press and AeroPress belong to machine filtration.

(   ) 3. If you use a siphon (or syphon) coffeemaker, you need to brew the coffee with artificial pressure to separate the coffee grounds and the liquid.

(   ) 4. The siphon (or syphon) coffeemaker makes use of the pressure difference caused by steam cooling down to filter coffee.

(   ) 5. The Moka pot brews coffee by passing boiling water pressurised by steam through ground coffee.

Situation 2

Step 1   Role-playing

Answers to the Tasks

A: Now let's use the semi-automatic espresso machine to make some coffee.

B: Great. Please show me how to make a cup of double espresso.

A: OK. The first part is about cleaning. Now take out a clean and dry cloth. Hang it on the ring of your apron, like this. Use one corner of the cloth to wipe the basket of the portafilter clean.

B: Done. How about the next step?

A: This is the control panel. Switch it on to connect to the power supply. Now put the cups under the brewing heads. Then press the Extraction Control to activate the flow of hot water through the group heads, flowing into the cups. The heads and the cups will be cleaned with the hot water. Watch out for hot water.

B: Am I right?

A: Well done. Now put the cups on the cup warmer like this. Next, clean the steam wand. Turn on the hot water control to clean the steam nozzle with steam, and drain the milk dirt and cold water left inside. Be careful not to be scalded by steam during operation.

B: OK.

A: Take it easy. Do it step by step. Very nice. Now let's go to the second part: to make a cup of double espresso.

B: Bravo.

A: We dose 14 g of the coffee grounds into the basket. Then distribute the grounds evenly by gently shaking the portafilter or by hand. You can also use the distribution tool.

B: Do you mean this wooden strip?

A: Yes. Then press the grounds with tamper. Try to make an even bed of grounds.

B: I'll try.

A: OK, now press the Extraction Control to drain the water and remove the coffee grounds from the brewing head, and lower the temperature of the brewing head. Note that the water is very hot.

B: Thank you for your reminder.

A: You're welcome. Take a closer look at the next steps. First insert and fasten the portafilter into the brewing head. Press the Extraction Control for Two Shots of Espresso, and then take the cup from the cup warmer and place it under the portafilter spout as soon as possible. After about 5 seconds, the coffee starts to flow into the cup. When the extraction time is up, the coffee will stop flowing automatically. Now it's done.

A: It smells so good. My barista, I can't wait to try it immediately. Could I?

B: Of course. Take care. It's still very hot. How do you think?

A: I like it very much. Strong fruit flavor. A little sweet with a little acidic but not bitter at all.

B: I'm glad you like it. You know I've experimented it for thousands of times.

A: It's all worth it. Anything else should I know?

B: Yes. We still need to clean the machine at once. Three steps. Step one, remove the portafilter from the brewing head and clean it with a clean dry cloth. Then put it upside down on the cup warmer. Step two, flush the brewing head with some hot water and rinse off the spouts at the same time. Make sure there is no dregs left. Step three, clean the cups and put them upside down on the cup warmer.

A: Could we use the espresso to make cold coffee?

B: Of course we can. But if you want to make cold coffee with chilled water, it would be better to use a French Press, a large pitcher or jar with towel-lined mesh sieve. The cold coffee will be made by a lager water ratio than regular hot coffee. And the coarsely grounds will be a best choice as the cold coffee will be set in a fridge for 12 to 24 hours.

A: Interesting. But for this time I wanna have a try to do the whole procedure of double espresso brewing by myself. Could I?

B: Yes. I'll help you when it is necessary.

A: Thank you very much.

 **Word and Phrase Bank**

New Words

basket /'bɑːskɪt/ *n.*   篓，筐
distribution /ˌdɪstrɪ'bjuːʃn/ *n.*   分发
barista /bə'riːstə; bə'rɪstə/ *n.*   咖啡师
regular /'regjələ(r)/ *adj.*   有规律的
procedure /prə'siːdʒə(r)/ *n.*   程序

Phrases & Expressions

power supply   供电电源

Extraction Control　萃取键

as soon as possible　尽快

have a try　尝试

Step 2　Consolidation Training

Ⅰ　Complete the following sentences with the words in the box. Change the form if necessary.

> same　machine　out　distribute　experimented

1. Now let's use the semi-automatic espresso _____ to make some coffee.

2. The heads will be cleaned and the previously retained cold water will be flushed _____.

3. We dose 14 g of the coffee grounds into the basket. Then _____ the grounds evenly by gently shaking the portafilter or by your hand.

4. You know I've _____ it for thousands of times.

5. Flush the group head with some hot water and rinse off the spouts at the _____ time.

Ⅱ　There are ten English words and terms numbered 1 through 10 in the left column. You should match them with their Chinese equivalents marked A through J in the right column and write down the corresponding letter in the right blank.

1. power supply	A. 程序
2. extraction control	B. 有规律的
3. procedure	C. 供电电源
4. have a try	D. 成千上万次
5. as soon as possible	E. 冲洗
6. regular	F. 尝试
7. thousands of times	G. 温杯器
8. rinse off	H. 稳定温度
9. cup warmer	I. 尽快
10. stabilizing the temperature	J. 萃取键

1. _____　2. _____　3. _____　4. _____　5. _____　6. _____

7. _____　8. _____　9. _____　10. _____

Answers to the Tasks

Task Module 4　Extensive Reading

Learn More about Brewing Espresso

How can we make a cup of perfect espresso? Some connoisseurs suggest that a well-brewed espresso should have a smooth layer of crema on the surface. The crema should be in a golden brown color, with fine surface and no large bubbles and spots. Although some people believe that the espresso could be brewed with coffee beans of various roast degrees, single or blend. But the most recommended coarseness is of the fine degree which allows the water to evenly penetrate through the grounds under high pressure and extract soluble substance. The process concentrates the balanced flavor between sweet and acidic, and the texture being smooth and creamy as well as a pleasant lingering aftertaste. Different people like different flavor and texture. Try to take more practice and find out what you like. If necessary, use an electric scale to measure the proportions.

If you want to have a cup of coffee with low caffeine, espresso is not a bad choice as it is often served in a smaller volume. Although the actual caffeine content of any coffee beverage varies by roast method, serving size, bean origin and variety, brewing method and other factors, a lot of research has proved that the above results are scientific.

Espresso is not only a drink that can be served alone, but also the basis of many kinds of coffee drinks. You could get more information about them from the next task Coffee Drinks.

Task Module 5　Creative Task

In order to produce more professional espresso based drinks, we need to know the ratios of different ingredients. Work in pairs. Try to find out the composition proportion of other espresso drinks. You are encouraged to collect more information from the library or through the internet if necessary.

Task 3　Coffee Drinks

Part One　Task Introduction

1. Teaching Objectives

Knowledge Objectives	1. Learn about the origin and development of coffee drinks. 2. Learn to classify the coffee drinks. 3. Get to learn the composition of the coffee drinks.
Skill Objectives	1. Be able to recognize the different types of coffee drinks. 2. In the process of learning English, feel the way of thinking in English, and improve the logic, speculation and innovation of thinking. 3. Be able to use English to understand and express information, views and feelings more accurately, and conduct effective cross-cultural communication.
Competency Objectives	1. Cultivate the sense of collaboration and teamwork. 2. Broaden the international vision, respect the diversity of the world culture, and build a sense of community with a shared future for mankind. 3. Build confidence and set up a correct view of English learning. 4. Deepen the understanding of Chinese culture and enhance cultural self-confidence through cultural comparison.

2. Teaching Focuses

(1) Get to learn the composition of the coffee drinks.

(2) Be able to recognize the different types of coffee drinks.

3. Teaching Difficulty

Effective intercultural communication in English.

Part Two　Task Implementation

Task Module 1　Pre-tasks

Step 1　Group Cooperation

Work in pairs. Discuss the professional terms in the box and get familiar with them. You

 咖 啡（双语）

can look them up online with your smartphones if necessary.

| drink beverage espresso brew cream foam caffeine decaffeinated |

Step 2　Panel Discussion

Work in pairs to discuss the following issues and exchange views.

1. What kinds of coffee drinks have you ever drunk in a café?

2. Do you know how to make the coffee beverage you like?

Task Module 2　Reading and Training
Step 1　Reading

Coffee Drinks

It is believed that the origin of coffee is Ethiopia. But according to the data, the word coffee originated from the Arabic qahwah and then in 1582 entered the English language via the Dutch koffie. And the earliest credible evidence of coffee-drinking as the modern beverage could be traced back to the middle of the 15th century in Sufi shrine. After so many years of development, coffee has become one of the world's major drinks and is immensely popular all over the world. The coffee drinks are becoming richer.

There are many ways to classify coffee drinks. According to the gears used to make coffees, there are espresso brewed by espresso machine and coffee brewed by Turkish Ibrik, French press, cloth brewer, Aeropress, syphon, stove-top top, chemex, etc. Espresso is usually stronger than most coffees made by other methods. It is the base for lots of other coffee drinks, like Cappuccino, Caffè Latte, Macchiato, Americano, etc.

If divided according to the composition of the drinks, coffee can be separated into pure coffee and coffee combinations. As for pure coffee drinks, they contain only coffee without any supplements, while coffee combinations could be made by coffee with milk, cream, condensed milk, honey, syrup, tea, liqueur, cocoa, cinnamon and so on. You can get some examples from the following table.

Names of combined coffee drinks	Ingredients
Americano	espresso and hot water
Macchiato	espresso and milk foam
Con Panna	espresso and whipped cream
Affogato	espresso and vanilla ice cream

Continuation table

Names of combined coffee drinks	Ingredients
Irish Coffee	strong brewed coffee, brown sugar, Irish whiskey and whipped cream
Turkish Coffee	water, sugar, superfine-grind coffee, cardamom, cinnamon, or nutmeg (if desired)

According to the coffee powder used to brew coffee, there are conventional coffee, instant coffee and coffee made by the single-serve containers. Conventional coffee drinks are brewed with the powder directly ground by coffee beans. Instant coffee drinks are made by the dehydrated powder or concentrated liquid through various manufacturing processes. Generally speaking, the instant coffee could be the mixture of a concentrated coffee with or without milk and/or sugar. Coffee made by the single-serve containers is the coffee drink brewed with coffee bags, pods or capsules. A container with the pre-packaged coffee in it is very convenient to use but meanwhile becomes one of the worst forms of human waste.

Coffee could also be sorted into caffeinated and decaffeinated coffees based on the caffeine content.

Word and Phrase Bank

New Words

origin /ˈɒrɪdʒɪn/ *n.*　起源，（原）产地

Ethiopia /ˌiːθiˈəʊpiə/ *n.*　埃塞俄比亚

Arabic /ˈærəbɪk/ *adj.*　阿拉伯的

Dutch /dʌtʃ/ *n.*　荷兰语

credible /ˈkredəb(ə)l/ *adj.*　可信的，可靠的

shrine /ʃraɪn/ *n.*　神殿

immensely /ɪˈmensli/ adv.　极其，非常

rich /rɪtʃ/ *adj.*　丰富的

classify /ˈklæsɪfaɪ/ *v.*　把……分类

gear /ɡɪə/ *n.*　装备，装置

espresso /eˈspresəʊ/ *n.*　（用蒸汽加压煮出的）浓缩咖啡

brew /bruː/ *v.*　冲（咖啡），沏（茶）

divide /dɪˈvaɪd/ *v.*　划分

composition /ˌkɒmpəˈzɪʃ(ə)n/ *n.*　成分构成，成分

supplements /ˈsʌplɪmənt/ *n.*　添加物

cream /kri:m/ *n.* 奶油

syrup /'sɪrəp/ *n.* 糖浆

liqueur /lɪ'kjʊə(r)/ *n.* 利口酒

cocoa /'kəʊkəʊ/ *n.* 可可粉

cinnamon /'sɪnəmən/ *n.* 肉桂

vanilla /və'nɪlə/ *n.* 香草 *adj.* 香草味的

Irish /'aɪrɪʃ/ *n.* 爱尔兰人；爱尔兰语；爱尔兰的

whiskey /'wɪski/ *n.* 威士忌酒

cardamom /'kɑ:dəməm/ *n.* 小豆蔻

nutmeg /'nʌtmeg/ *n.* 肉豆蔻

powder /'paʊdə(r)/ *n.* 粉，粉末

grind /graɪnd/ *v.* 碾磨

dehydrated/ˌdi:haɪ'dreɪtɪd/ *adj.* 脱水的

concentrated /'kɒnsntreɪtɪd/ *adj.* 浓缩的

mixture /'mɪkstʃə(r)/ *n.* 混合物

pod /pɒd/ *n.* 包

capsule /'kæpsju:l/ *n.* 胶囊

sort /sɔ:t/ *v.* 把……分类

caffeinated /'kæfɪneɪtɪd/ *adj.* 含咖啡因的

decaffeinated /di:'kæfɪneɪtɪd/ *adj.* 脱咖啡因的

caffeine /'kæfi:n/ *n.* 咖啡因

Phrases & Expressions

trace back   追溯

Turkish Ibrik   土耳其伊布里克咖啡壶

French press   法压壶

cloth brewer   布袋滴漏

Aeropress   爱乐压壶

syphon   虹吸壶

stove-top top   炉顶摩卡壶

chemex   手冲滤壶

Cappuccino   卡布奇诺

Caffè Latte   拿铁

Macchiato   玛奇朵

Americano   美式

coffee combinations   花式咖啡

condensed milk   炼乳

milk foam   牛奶奶泡

Con Panna　康宝蓝咖啡

whipped cream　稀奶油

Affogato　阿芙佳朵咖啡

strong brewed coffee　浓咖啡

brown sugar　红糖

superfine-grind coffee　超细研磨咖啡

conventional coffee　传统咖啡

instant coffee　速溶咖啡

single-serve containers　单杯咖啡

manufacturing processes　生产工艺

Step 2　After-reading Training

I　Answer the questions.

1. How many classifications of coffee drinks does the author introduce in the article?

_____

2. What coffee gears are mentioned in the article?

_____

3. If divided according to the composition of the drinks, what types can the coffee be separated into?

_____

4. What are compositions in the Irish Coffee?

_____

5. What are instant coffee drinks made of?

_____

II　Read the statements and tick the correct boxes.

Statements	Right	Wrong	Not mentioned
Espresso is usually stronger than most coffees made by other methods.			
If divided according to the composition of the drinks, coffee can be separated into pure coffee and instant coffee.			
Macchiato's composition includes espresso and whipped cream.			
Pure coffee drinks contain only coffee without any supplements.			

Continuation table

Statements	Right	Wrong	Not mentioned
Coffee could be sorted into caffeinated and decaffeinated coffees based on the caffeine content.			

Task Module 3　Communication Tasks

Answers to the Tasks

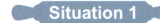

Situation 1

Step 1　Role-playing

A: Hi, Evan. Nice to meet you.

B: Hi, Justin. Nice to meet you again. Yesterday I went to Café de La Mode. I was totally confused about the drinks. Could you tell me how you would classify the coffee drinks?

A: In my opinion, we could sort the coffee drinks into 3 types. They're the original coffee, the classics and others. But for a Café, it will classify the drinks in another way.

B: Do some explanation of classification first. OK?

A: OK. Let's get started with the original coffee. The original coffee is coffee without any additives. We usually divide original coffee into single-original coffee and blended coffees. Both of them could be brewed by any gear.

B: Single-original coffee? Could you explain it in detail?

A: Single-original coffee is the coffee produced with coffee beans from a single production area and a single harvest season, like Yirgacheffe of Ethiopia, Blue Mountain Coffee of Jamaica, Mande Ling of Sumatra and Yunnan Arabica coffee of China.

B: So blended coffees are the mixtures of different coffees. Right?

A: I think so.

B: How about the classics?

A: I think the classics generally are the espresso-based coffee drinks adding with milk, cream, chocolate sauce or water, etc. For example, Cappuccino, Caffè Latte, Macchiato, Flat White and Café au lait include espresso and milk with different ratios. And Caffè Mocha is a classic combination of espresso, chocolate sauce and milk. You could use espresso, milk and single cream to make Breve while making Espresso Con Panna without milk. And Americano, Ristretto and Lungo are the drinks including espresso and water.

B: Cool. Then what about the other coffees?

A: The other coffees could be iced or hot with different additives like alcohol, yogurt, icecream, almond, etc. You could create different flavors by adding different additives.

B: It seems very interesting. Could you make a cup of alcoholic coffee for me to try?

A: No problem. Which one would you like, whisky, brandy or martini?

B: Martini, please. Thank you.

A: You're welcome. Please wait for a moment.

 Word and Phrase Bank

New Words

additive /'ædətɪv/ *n.*   添加物

blended /'blendɪd/ *adj.*   混合的

Jamaica /dʒə'meɪkə/ *n.*   牙买加

Sumatra /sʊ'mɑːtrə/ *n.*   苏门答腊岛

ratio /'reɪʃɪəʊ/ *n.*   比例

alcohol /'ælkəhɒl/ *n.*   含酒精饮品

yogurt /'jɒgət/ *n.*   酸奶

almond /'ɑːmənd/ *n.*   杏仁

flavor /'fleɪvə/ *n.*   风味

Phrases & Expressions

original coffee   原味咖啡

single-original coffee   单品咖啡

single production area   单一产区

Yirgacheffe   耶加雪菲咖啡

Blue Mountain Coffee   蓝山咖啡

Mande Ling   曼特宁咖啡

Arabica coffee   阿拉比卡咖啡

chocolate sauce   巧克力酱

Flat White   馥芮白

Café au lait   欧蕾

Caffè Mocha   摩卡咖啡

single cream   稀奶油

Breve   布雷卫咖啡

Ristretto   意大利超浓咖啡

Lungo   廊构咖啡

brandy   白兰地

martini   马提尼

Step 2　Consolidation Training

I　Match the words on the left with their meanings on the right.

1. additive	A. combined or mixed together
2. alcohol	B. a substance that is added in small amounts to something
3. blended	C. drinks such as beer, wine, etc. that can make people drunk
4. flavor	D. a relationship between two things when it is expressed in numbers or amounts
5. ratio	E. a particular type of taste

1. _____　2. _____　3. _____　4. _____　5. _____

II　Decide whether the following statements are true (T) or false (F).

(　　) 1. The original coffee is coffee without any additives.

(　　) 2. Single-original coffee is the coffee produced with coffee beans from a single production area and a single harvest season.

(　　) 3. Yirgacheffe of Ethiopia and Blue Mountain Coffee of Jamaica are not single-original coffee.

(　　) 4. Flat White and Café au lait are classic combination of espresso, chocolate sauce and milk.

(　　) 5. The other coffees could be iced or hot with different additives like alcohol, yogurt, icecream, almond, etc.

Situation 2

Step 1　Role-playing

Answers to the Tasks

A: Justin, what is specialty coffee?

B: It is a term referring to single-original coffee with outstanding flavor.

A: Is there a standard to define it?

B: Yes. The Specialty Coffee Association sets clear standards for scoring specialty coffee.

A: I'm curious about the standards of specialty coffee.

B: Simply speaking, if the cup test score of SCA is more than 80, the coffee could be called high-quality coffee.

A: Could you tell me more details about the scoring?

B: Yes. According to the standard, coffee cupping with a score higher than 80 are called boutique coffee. And coffee with a score of 80–84.99 is rated as "Very good", coffee with a score of 85–89.99 is rated as "Excellent", and coffee with a score of 90–100 is rated as "Outstanding".

A: I see. Where does the specialty coffee generally come from?

B: Specialty coffee is typically grown in three continents of the "Bean Belt or Coffee Belt", including South and Central Americas, Asia, and Africa.

A: What is the world's most expensive specialty coffee?

B: I'm not sure. But it's said that Geisha of Panama and Yirgacheffe of Ethiopia are the two most expensive specialty coffees.

A: I really want to try these best specialty coffees.

B: So do I.

Word and Phrase Bank

New Words

term /tɜːm/ *n.* 术语

outstanding /aʊt'stændɪŋ/ *adj.* 显著的, 突出的

standard /'stændəd/ *n.* 标准

define /dɪ'faɪn/ v. 给……下定义

association /əˌsəʊsi'eɪʃ(ə)n/ *n.* 协会

curious /'kjʊəriəs/ *adj.* 好奇的

scoring /'skɔːrɪŋ/ *n.* 评分

boutique /buː'tiːk/ *n.* 精品

rate /reɪt/ *v.* 评定

continent /'kɒntɪnənt/ *n.* 洲，大陆

Panama /'pænəmɑː/ *n.* 巴拿马

Phrases & Expressions

refer to   描述, 涉及，参考，指的是

specialty coffee   精品咖啡

cup test   杯测

coffee cupping   咖啡杯测

Geisha   瑰夏咖啡

Step 2   Consolidation Training

I   Complete the following sentences with the words in the box. Change the form if necessary.

association   cup test   define   generally   curious   refer to   outstanding
continent   scoring   term

1. "Darling" is a _____ of endearment.

2. In ancient times, people _____ a happy life as a life with food and shelter.

3. The _____ was formed to represent the interests of women artists.

4. I am _____ about what keeps him abroad.

5. It's a great help for _____ well in the exam.

6. He traversed alone the whole _____ of Africa from east to west.

7. Writers often _____ a dictionary.

8. If the _____ score of SCA is more than 80, the coffee could be called high-quality coffee.

9. _____ speaking, the more you pay, the more you get.

10. Her _____ ability earned her a place in the company.

Ⅱ   There are ten English words and terms numbered 1 through 10 in the left column. You should match them with their Chinese equivalents marked A through J in the right column and write down the corresponding letter in the right blank.

1. outstanding	A. 标准
2. standard	B. 精品
3. boutique	C. 巴拿马
4. rate	D. 显著的, 突出的
5. Panama	E. 评定
6. specialty coffee	F. 瑰夏咖啡
7. coffee cupping	G. 评分
8. Geisha	H. 协会
9. scoring	I. 精品咖啡
10. association	J. 咖啡杯测

1. _____   2. _____   3. _____   4. _____   5. _____   6. _____

7. _____   8. _____   9. _____   10. _____

Task Module 4   Extensive Reading

Answers to the Tasks

Learn More about the Coffee Drinks

In Italy, the cappuccino is traditionally drunk in the morning, usually as part of breakfast and only consumed up to 11:00 a.m., since cappuccinos are milk-based and regarded too heavy to drink later in the day. The cappuccino is widely considered to be the coffee drink with the most harmonious ratio of coffee and milk, which makes it one of the most popular coffee beverages on the basis of espresso. Here is the classic formulation of cappuccino.

1) Equipment you need:

2 medium cups

Espresso machine

Milk pitcher

2) Ingredients:

16-20 g fine ground coffee

130-150 mL cold and fresh milk

Chocolate or cinnamon powder, optional

3) Serves for 2 cups.

4) Temperature: hot

5) Making process

Step 1: Heat the cup with hot water or put the cup on the coffee machine.

Step 2: Brew a cup (25 mL) of espresso in each cup.

Step 3: Heat the milk for about 20 seconds to make fine milk foams.

Step 4: Pour the milk on each cup of espresso, and let the foam float on the coffee surface. In order to produce strong coffee flavor in the first sip of coffee, try to keep a circle of coffee crema around the cup.

Step 5: According to your preference, you can sprinkle some chocolate powder or cinnamon powder on the foam with a vibrator or a sieve.

Task Module 5　Creative Task

Here are two tasks. One is to help people more easily to understand the history of specialty coffee. The other one is to help people know more about the recipes of different coffee drinks. You and your partner could select any of them. For the first one, please make a report to show your idea. For the second one, please collect information about some coffee drinks you like. You can take the following table as an example. Then share your report or recipes with others. You are encouraged to collect more information from the library or through the internet if necessary.

Names of drinks	Ingredients and recommended amount
Doppio	2 shots/50 mL of espresso
Caffè Latte	1 shot/25 mL of espresso and steamed milk (with about 5 mm layer of foam)
Con Panna	2 shots/50 mL of espresso and 1 tablespoon of whipped cream
Caffè Mocha	4 tablespoons of dark chocolate sauce, steamed milk (with about 1 cm layer of foam), 2 shots/50 mL of espresso

项目 3

咖啡冲泡

任务1　咖啡杯测

 ## 第1部分　任务介绍

咖啡杯测

1. 教学目标

知识目标	1. 了解咖啡杯测的发展情况。 2. 掌握杯测的类型和作用。 3. 通过学习，理解杯测的专业标准，掌握风味轮和感官词汇中必要的专业知识。
技能目标	1. 能够区分杯测所需器具，通过操作杯测过程展示其用途。 2. 在学习英语的过程中，感受英语思维方式，提升思维的逻辑性、思辨性和创新性。 3. 能够运用英语，比较准确地理解和表达信息、观点、情感，能有效地进行跨文化交际与沟通。
素质目标	1. 培养与人协作的精神和团队合作的意识。 2. 学习感受风味轮和感官词汇中的专业规范，培养严谨、细致、标准化的专业精神和职业素养。 3. 学习杯测师或品鉴师实事求是的专业态度以及勤学苦练、精益求精的工匠精神。 4. 学会选择学习资源，能运用恰当的英语学习策略进行自主学习。

2. 教学重点

（1）咖啡杯测的器具和用途。

（2）咖啡杯测的标准化操作流程。

3. 教学难点

能用英语进行跨文化交际与沟通。

第2部分 任务实施

任务模块1 前置任务

步骤1 小组合作

两人一组，讨论下列方框中的专业用语并加以熟悉。如果需要，你可以使用智能手机在线查找。

> cupping tasting cupper G-grader SCA evaluate defective bean

步骤2 小组讨论

两人一组，讨论以下问题并交流观点。

1. 人们如何评价咖啡的质量？
2. 人们在评价咖啡时会使用什么样的标准？

任务模块2 阅读与训练

咖啡杯测

咖啡杯测或咖啡品鉴是一种评估咖啡的方法。根据咖啡从农场到成为杯中饮品的整个产业链，有产地杯测、批量销售咖啡生豆的杯测和烘焙好的咖啡熟豆的杯测。

产地杯测是最基本的杯测方法。咖啡寻豆师经常使用这种方法帮助他们确定这些咖啡豆的价格，并根据不同产区进行评估。这个阶段的杯测，最重要的是找出这些豆子的风味特征，并检查这些豆子是否受到污染。

杯测后，咖啡豆将按照不同的标准进行分级，如埃塞俄比亚的缺陷咖啡豆比例分级标准（G1-G5），肯尼亚的粒度分级标准（E、AA、AB、C、PB、TT、T、MH/ML）等。分级后，咖啡将根据不同的等级以不同的价格出售。对于这一阶段，结果主要与瑕疵率、大小、生长海拔、硬度等有关，与咖啡风味无关。烘焙好的咖啡豆的杯测是通过品尝和鉴别咖啡来评估其特性的做法，将有助于决定烘焙曲线、质量控制和购买决策。

据说杯测这种做法已被使用了100多年，可以追溯到19世纪末，当时商家在购买过程中为确保品质统一而用这种方法来做检测。20世纪末，"超凡杯"比赛中开始使用杯测来判断和评鉴咖啡样品后，这种方法被广泛使用。后来，美国精品咖啡协会（SCA）制定的杯测协议成为一项流行的准则，至今仍是咖啡社区的标准指南。

杯测是咖啡品质学会的咖啡品质鉴定师的专业行为。但相对简单的技术也可以由任何人非正式地完成。一些专家认为，杯测是基于感官评估的，是主观的，其结果是否准确将受到品尝者能力的影响。因此，他们建议将人员操作的杯测（严格按照专业人士

制定的SCA协议执行）与技术（如近红外光谱和人工智能）结合起来，以产生可靠的杯测结果。

任务模块3　场景交际任务

A：娜塔莉，你在干什么？

B：我在准备做咖啡杯测。

A：为什么要做这个？

B：我们想知道昨天烘焙的咖啡豆的烘焙效果。

A：这些是做什么用的？

B：这是水，温度通常在92.2～94.4 ℃。

A：这是什么？

B：这是磨豆机。我们将使用中等粒度的咖啡样品进行杯测，以便最大限度地进行咖啡萃取。

A：这些杯子是用来做什么的？

B：我们称这些为杯测杯。尽管SCA建议使用最大容量为220毫升的标准杯测杯，但我认为任何容量为110～300毫升的杯子都可以。有两点需要注意。一是咖啡与水的比例。对于我们的杯子，我们使用8.25克咖啡豆和150毫升水。对于220毫升的杯子，12克咖啡豆将是完美的选择。二是使用带盖的杯子或碗。当你把它盖在杯子或碗上时，它可以帮助保持磨碎后的咖啡的香味。

A：其他的呢？

B：这是杯测勺。它将被用来破开咖啡渣。

A：什么是"咖啡渣"？

B：我们在咖啡中加入热水后，很快就会有一层咖啡颗粒漂浮在表面，我们称为"咖啡渣"。

A：我明白了。

B：我们还用勺子撇去表面上多余的泡沫。

A：还有什么？

B：这个装满水的大玻璃杯是用来清洗勺子的。还有一个数字秤、一个计时器和一个记事本。

A：你为什么需要记事本？

B：看，这里有一些杯测的表格。杯测师需要详细记录咖啡品质的各个方面，主要包括干（湿）香气、风味、酸质、体脂感、一致性、干净度、平衡性、甜度等级、优缺点和综合考虑等。

A：哇，这对我来说实在太复杂了。

A：娜塔莉，怎么做杯测？

B：让我做一杯给你看。第一步是研磨咖啡样品。称出8.25克咖啡豆，然后将磨好的咖啡粉倒入这个杯子。

A：下一步怎么做？

B：现在我们闻一闻磨碎的干咖啡粉并评估它们。我们弯下腰去闻。记住，不要用手触摸杯子。

A：为什么？

B：你手上的烟味、护手霜味或肥皂味会留在杯子上。这将影响我们的识别和判断。

A：这是否意味着应该避免香水、化妆品或香烟之类的气味？

B：完全正确。我们需要在杯测表上做记录。然后，我们将向杯子里倒入94 ℃的水。为了确保咖啡粉完全浸透而不结块，用大流量把水倒满杯子。咖啡粉与水的比例一般为1：18.18。

A：你的操作很熟练。

B：谢谢你。当我们开始倒水时，同时开始计时。最好的方法是用计时器倒数4分钟。时间到了，它会发出嘟嘟声提醒你。

A：好的。我现在该怎么办？

B：让我们等等。1分钟后，我们来闻咖啡的香味。小心，它们很烫。然后在杯测表上写下你的感受。

A：明白了。哇，这个闻起来很香。

B：咖啡渣将在4分钟内形成。当计时器发出哔哔声时，我们将一起来破渣。

A：如何破渣？

B：像这样，分3次破渣。然后闻它的芳香。很好。下一步是同时用两个杯勺撇去泡沫并拖动。记住，在移动到下一个杯子之前，你应该冲洗勺子，并点击勺子尖上的纸。不允许交叉污染。

A：好的。冲洗，点击，下一杯。我会了。我什么时候可以品尝咖啡？

B：注水后8～10分钟，当温度降到50～60 ℃时，评测就开始了。现在用勺子舀起液体到嘴里，快速地发出咕噜声，以尽可能覆盖口腔的大部分区域。

A：我应该特别注意什么？

B：咖啡的体脂感、酸质、甜度和风味。如有必要，你可以重复这些步骤。最后一步是做笔记。

A：这似乎并不难，但要准确地做到这一点并不容易。

B：是的。为了准确，专业的杯测师或品鉴师不仅需要敏锐的感官，而且需要大量的练习。

任务模块4 知识拓展

了解更多关于咖啡杯测的信息

咖啡被认为是市场上化学成分最复杂的饮料之一。咖啡的香气、质地和风味很难与其他食物相比。有很多因素会影响咖啡的品质，比如咖啡籽的基因、生长地点和生长方式、处理、储存、运输、烘焙和萃取等。因此，为了找到我们想要的咖啡，我们需要尝试闻、喝、尝、感受和评价咖啡。杯测师或品鉴师应运而生。

为了建立咖啡师比赛的标准，为其他人评测咖啡提供参考，泰德·林格于20世纪90年代末为美国精品咖啡协会（SCA）创建了咖啡品鉴师风味轮。为了创造咖啡感官质量的通用语言和测量工具，世界咖啡研究所组织了一个由堪萨斯州立大学10名感官科学家组成的小组，利用世界顶级感官科学中心之一的堪萨斯州立大学感官分析中心，对咖啡味道进行重复性研究。2016年，世界咖啡研究感官词汇发展到110个属性并出版。随着咖啡新品种的不断培育，上述文献被称为活档案。多亏了这些文献，咖啡杯测和品鉴才得以更专业，更易于描述，更可靠。

任务模块5 创意任务

小组合作。试着找出《咖啡品鉴师风味轮》或《世界咖啡研究感官词典》的文献。学习如何使用它们，并与他人分享你的学习成果。也许有很多新单词和表达。别担心。如有必要，鼓励你从图书馆或通过互联网收集更多信息。

任务模块6 任务执行能力考核

序号	考核细分项目	标准分数（分）	得分（分）
1	专业用语	10	
2	任务资料的基础训练	20	
3	场景任务1专项训练	20	
4	场景任务2专项训练	20	
5	知识拓展	10	
6	创意任务	20	

任务模块7 目标达成考核

评分考核项目	标准分数（分）	个人自评	小组互评	教师评分
理论知识	30			
技能训练	40			
职业素养	30			
总分（分）	100			
综合总分				
说明	综合总分=个人自评（占总分的20%）+小组互评（占总分的20%）+教师评分（占总分的60%）			

任务2　咖啡冲煮

 ## 第1部分　任务介绍

咖啡冲煮

1. 教学目标

知识目标	1. 了解不同的咖啡冲煮方法和冲煮咖啡时应考虑的因素。 2. 掌握常见咖啡冲煮设备的特点和基本原理。 3. 掌握意式浓缩咖啡的特点和常见浓缩咖啡饮品的成分。 4. 能够识别出半自动浓缩咖啡机的主要部件及其功能。
技能目标	1. 能够理论联系实践，正确操作半自动浓缩咖啡机。 2. 在学习英语的过程中，感受英语思维方式，提升自身思维的逻辑性、思辨性和创新性。 3. 能够运用英语，比较准确地理解和表达信息、观点、情感，能有效地进行跨文化交际与沟通。
素质目标	1. 培养与人协作的精神和团队合作意识。 2. 通过半自动浓缩咖啡机的操作训练，培养咖啡师时刻牢记安全操作、严谨规范、专注细心的工匠精神。 3. 学习不同的咖啡饮品的成分配方，培养积极创新、追求品质、追求突破的创新意识。 4. 树立信心，不畏困难，积极探索，树立正确的英语学习观。

2. 教学重点

（1）常见咖啡冲煮设备的特点和区别。

（2）半自动浓缩咖啡机的规范操作。

（3）咖啡饮品的特点与创新。

3. 教学难点

能用英语进行跨文化交际与沟通。

第2部分　任务实施

任务模块1　前置任务

步骤1　小组合作

两人一组，讨论下列方框中的专业用语并加以熟悉。如果需要，可以使用智能手机在线查找。

> coffeemaker　beverage　decoction　steep　pressurization　extraction　filter

步骤2　小组讨论

两人一组，讨论以下问题并交流观点。

1. 你知道哪几种咖啡机？它们之间有什么区别？
2. 你知道如何冲煮一杯咖啡吗？你自己做过吗？

任务模块2　阅读与训练

咖啡冲煮

咖啡冲煮是将咖啡粉制成饮品的过程。有人建议咖啡应在饮用前冲煮。不同地区的人可能用不同的方法冲煮咖啡。例如，土耳其咖啡也称阿拉伯咖啡，被认为是当今咖啡的雏形。土耳其咖啡是将研磨极细的咖啡粉末通过烹煮制成的咖啡饮品。土耳其咖啡在土耳其、希腊和巴尔干半岛很受欢迎。印度过滤咖啡在印度南部尤为常见。

煮沸或煎煮是用冷水煮咖啡粉的方法。对于土耳其咖啡而言，只会煮研磨得非常细的咖啡。咖啡煮沸3次，表面出现深棕色的浓密泡沫，底部为咖啡粉末。早期的"牛仔咖啡"是在野外用罐子或罐子里的冷水加热粗糙的咖啡渣通过煮沸的方法煮出来的，饮用前加糖或香料，以掩盖其苦涩的味道。

冲煮被认为是一种相对传统、自然、简单的方式。冲煮可以进一步分为人工压滤、真空冲泡、滴滤和摩卡壶蒸汽加压。通常在约1个大气压下，用94 ℃（201 °F）左右的热水冲煮咖啡粉。加压特指意大利浓缩咖啡机，它通过在约9个大气压下将92～96 ℃（198～205 °F）的热水穿过咖啡粉中，在大约25秒钟内煮出咖啡。这两种方法都需要将咖啡粉与热水混合，并在有限的时间内浸泡，以获得极佳的风味，且尽可能少的苦味。咖啡液中没有咖啡渣。据说，咖啡冲泡和饮用的相对理想温度为：温度在92～96 ℃（198～205 °F）萃取咖啡粉；在85～93 ℃（185～199 °F）浸泡咖啡粉；在68～79 ℃（154～174 °F）饮用咖啡。

冲泡咖啡时，通常考虑以下因素：咖啡粉量、咖啡粉的研磨度、咖啡粉粗细的均匀性、水温、粉水比例（咖啡粉与水的比例）、咖啡粉的萃取方式（如何用水萃取风味）、所需的调味品（如糖、牛奶和香料）以及分离技术（分离液体和废咖啡渣）等。

任务模块3　场景交际任务

场景1

A：早上好，亚历克斯。哇，这些包裹里都是什么？

B：它们是不同的咖啡机。你能帮我把它们拆开吗？

A：没问题。最大的这个包裹是什么？

B：这是一台商用半自动咖啡机。

A：那些呢？

B：它们是手冲咖啡壶（又称美式滤泡壶）、法压壶、爱乐压咖啡壶、虹吸式咖啡壶和……

A：等等。我完全糊涂了。

B：让我换种方式跟你说。一般来说，咖啡机主要有两种类型。第一种是这样的，意大利浓缩咖啡机，它可以是半自动和全自动的。

A：它们之间有什么区别？

B：使用全自动咖啡机时，你不需要手动研磨和压实咖啡。它还会自动萃取浓缩咖啡。但对于半自动的咖啡机来说，前两步必须由你动手完成。你还需要一台额外的咖啡研磨机来研磨咖啡粉。

A：我明白了。另一种是怎样的？

B：另一种我们称为咖啡壶。咖啡壶是通过煮沸或浸泡来萃取咖啡的。人们通常又将浸泡萃取细分为人工压滤、真空冲泡、滴滤和蒸汽加压。

A：你能给我举几个例子吗？

B：可以。这是法压壶和爱乐压咖啡壶，属于人工压力过滤。你需要用人工压力来萃取咖啡，以分离咖啡渣和咖啡液体。这意味着你需要像这样用手按压它。

A：明白了。

B：这个是虹吸式咖啡壶。它利用蒸汽冷却产生的压力差过滤咖啡，它是一台真空冲泡式咖啡壶。

A：这个带杯子的是什么？

B：这是Hario V60。Hario V60和这款Chemex手冲咖啡壶都属于手冲咖啡壶，这是利用自然重力来滴滤萃取咖啡的。所以把它们归类到滴滤式咖啡壶中。

A：现在介绍下这款银色的吧。

B：这是一个摩卡咖啡壶，通过将蒸汽加压的沸水浸透咖啡粉来冲煮咖啡。所以把它归类为蒸汽加压咖啡壶。

A：我基本上理解了。你能告诉我怎么用吗？

B：这是我的荣幸。不过首先，我们需要把它们清洁干净，并找地方把它们放好。

场景2

A：现在让我们用这款半自动浓缩咖啡机来萃取咖啡。

B：太好了。请告诉我如何做一杯双份意式浓缩咖啡。

A：好的。第一部分是清洁工作。拿出一块干净的干布，像这样把它挂在你围裙的环上。用布的一角将手柄和粉碗擦干净。

B：完成。下一步怎么做？

A：这是控制面板。打开开关连接电源。现在把杯子放在冲煮组头下方。按下萃取键，让热水流出冲煮组头，流入杯子里，冲煮组头和杯子就可以用热水洗干净了。小心热水。

B：我这样做对吗？

A：很好。现在像这样把杯子放在暖杯盘上。接下来要清洁蒸汽棒。打开这个蒸汽开关，用蒸汽清洗蒸汽喷嘴，把里面残留的牛奶垢和冷水排掉。操作时，注意不要被蒸汽烫到了。

B：好的。

A：别紧张。一步一步地做。非常好。现在进入第二部分：做一杯双份浓缩咖啡。

B：太棒了。

A：我们把14克咖啡粉装入手柄粉碗里。然后，轻轻摇晃手柄或用手均匀地抹匀咖啡粉。你还可以使用布粉工具。

B：你说的是这个木条吗？

A：是的。接着用粉锤压粉。试着把粉压平、压匀。

B：我试试。

A：好，现在按萃取键放水，把冲煮头上的咖啡渣滓清除掉。同时，把冲煮头的温度降下来。注意水很烫。

B：谢谢你的提醒。

A：不客气。接下来的步骤你仔细看。先将手柄扣紧在冲煮头上，按下两杯浓缩咖啡萃取键，再尽快从暖杯盘上取下杯子放在手柄壶嘴下面。大约5秒钟后，咖啡开始缓缓流入杯子里。萃取时间到了，咖啡会自动停止流出。完成。

A：它闻起来真香。我的咖啡师，我迫不及待地想马上尝尝，可以吗？

B：当然可以。小心，现在还很烫。你觉得怎么样？

A：我非常喜欢。浓浓的果香味，有点甜，有点酸，但一点也不苦。

B：我很高兴你喜欢它。你知道我已经做了数千次试验。

A：这一切都值得。还有什么我应该知道的吗？

B：是的。我们还需要马上清洗机器。清洗机器分为3个步骤。第一步，从萃取头上取下过滤器，并用干布清洁，然后将它倒置在杯托上。第二步，用热水冲洗组头，同时冲洗喷嘴，确保没有剩余的残渣。第三步，清洁杯子，将它倒置放在温杯器上。

A：我们可以用浓咖啡冲冷咖啡吗？

B：当然可以。但是，如果你想用冷冻的水来冲冷咖啡，最好用法压壶、大水罐或装有毛巾内衬网筛的罐子。制作冷咖啡时，水的比例比普通热咖啡更大。粗磨的咖啡将是最好的选择，因为冷咖啡将在冰箱中放置12～24小时。

  A：有趣。但这一次，我想尝试一下自己冲双份浓缩咖啡的整个过程。可以吗？

  B：可以。必要时我会帮助你。

  A：非常感谢。

任务模块4　知识拓展

了解更多关于冲煮意式浓缩咖啡的信息

  我们怎样才能冲出一杯完美的意式浓缩咖啡？一些鉴赏家建议，一杯精心酿造的浓缩咖啡表面应该有一层光滑的咖啡油脂。油脂应该是金棕色的，表面细腻，没有任何大气泡和斑点，颜色均匀。尽管有些人认为，浓缩咖啡可以用不同烘焙度的咖啡豆冲煮，无论是单一的还是混合的。但按粗糙度的话，最推荐的还是细粉，它允许水在高压下均匀渗透咖啡粉层，并萃取可溶性物质。该工艺可以使甜味和酸味达到平衡，口感顺滑，油脂余味悠长宜人。不同的人喜欢不同的味道和质地。试着多练习，找出你喜欢的。如有必要，使用电子秤测量比例。

  如果你想喝一杯低咖啡因的咖啡，浓缩咖啡是个不错的选择，因为它通常是小杯小量供应的。尽管任何咖啡饮品的实际咖啡因含量因烘焙方法、食用量、豆类来源和品种、冲煮方法和其他因素而异，但许多研究证明，上述结果是科学的。

  浓缩咖啡不仅是一种可以单独饮用的饮品，也是多种咖啡饮料的基础。你可以从下一个任务"咖啡配方"中获得更多此类信息。

任务模块5　创意任务

  为了生产更专业的浓缩咖啡饮料，我们需要知道不同成分的比例。两人一组，试着找出其他浓缩咖啡饮料的成分比例。如果需要，鼓励你从图书馆或通过互联网收集更多信息。

任务模块6　任务执行能力考核

序号	考核细分项目	标准分数（分）	得分（分）
1	专业用语	10	
2	任务资料的基础训练	20	
3	场景任务1专项训练	20	

序号	考核细分项目	标准分数（分）	得分（分）
4	场景任务2专项训练	20	
5	知识拓展	10	
6	创意任务	20	

任务模块7　目标达成考核

评分考核项目	标准分数（分）	个人自评	小组互评	教师评分
理论知识	30			
技能训练	40			
职业素养	30			
总分（分）	100			
综合总分				
说明	综合总分=个人自评（占总分的20%）+小组互评（占总分的20%）+教师评分（占总分的60%）			

任务3　咖啡配方

 第1部分　任务介绍

咖啡配方

1. 教学目标

知识目标	1. 了解咖啡饮品的发展情况。 2. 掌握咖啡饮品的类型和制作配方。
技能目标	1. 能够辨别出不同的咖啡饮品。 2. 在学习英语的过程中，感受英语思维方式，提升思维的逻辑性、思辨性和创新性。 3. 能够运用英语，比较准确地理解和表达信息、观点、情感，能有效地进行跨文化交际与沟通。
素质目标	1. 培养与人协作的精神和团队合作的意识。 2. 拓宽国际视野，尊重文化多样性，树立人类命运共同体意识。 3. 树立自信，树立正确的英语学习观。 4. 通过文化比较，加深对中国文化的理解，增强文化自信。

2. 教学重点

（1）掌握咖啡饮品的制作配方。

（2）能够辨别出不同的咖啡饮品。

3. 教学难点

能用英语进行有效的跨文化交际与沟通。

 第2部分　任务实施

任务模块1　前置任务

步骤1　小组合作

两人一组，讨论下列方框中的专业用语并加以熟悉。如果需要，你可以使用智能手机在线查找。

> drink　beverage　espresso　brew　cream　foam　caffeine　decaffeinated

步骤2　小组讨论

两人一组，讨论以下问题并交流观点。

1. 你在咖啡馆喝过什么咖啡？
2. 你知道如何制作你喜欢的咖啡饮品吗？

任务模块2　阅读与训练

咖啡配方

　　人们认为咖啡起源于埃塞俄比亚。数据显示，咖啡一词起源于阿拉伯语qahwah，之后于1582年通过荷兰语koffie进入英语。咖啡作为现代饮料最早的可靠证据可以追溯到15世纪中叶的苏菲神殿。经过这么多年发展，咖啡已经成为世界上主要的饮品之一，在全世界都非常受欢迎。咖啡饮品变得越来越丰富。

　　咖啡饮品有很多分类方法。根据冲煮咖啡所用的设备，可以分为用浓缩咖啡机冲泡的浓缩咖啡和用土耳其伊布里克、法压壶、布袋滴漏、爱乐压壶、虹吸壶、炉顶摩卡壶和手冲滤壶等冲煮的咖啡。浓缩咖啡通常比其他方法冲煮出来的大多数咖啡都要浓。它是配制许多咖啡饮品的基础，如卡布奇诺、拿铁、玛奇朵和美式咖啡等。

　　如果按照饮品的成分来划分，咖啡可以分为纯咖啡和花式咖啡。至于纯咖啡饮料，它们只含有咖啡，不含任何添加物。而花式咖啡可以由咖啡与牛奶、奶油、炼乳、蜂蜜、糖浆、茶、利口酒、可可、肉桂等混合制成。你可以从下表中获得一些例子。

花式咖啡饮品名称	成分
Americano 美式咖啡	浓缩咖啡和热水
Macchiato 玛奇朵咖啡	浓缩咖啡和牛奶奶泡
Con Panna 康宝蓝咖啡	浓缩咖啡和稀奶油
Affogato 阿芙佳朵咖啡	浓缩咖啡和香草冰激凌
Irish Coffee 爱尔兰咖啡	浓缩咖啡、红糖、爱尔兰威士忌和稀奶油
Turkish Coffee 土耳其咖啡	水、糖、超细研磨咖啡、小豆蔻、肉桂或肉豆蔻（如果需要）

　　根据用来冲泡咖啡的咖啡粉，可以分为传统咖啡、速溶咖啡和单杯咖啡。传统的咖啡饮品是用咖啡豆直接研磨的粉末冲煮的。速溶咖啡饮料是通过各种生产工艺制成的脱水粉末或浓缩液体。一般来说，速溶咖啡可以是浓缩咖啡加（或不加）牛奶、含（或不含）糖的混合物。单杯咖啡是用咖啡袋、咖啡包或咖啡胶囊封装的咖啡饮品。预先用容器包装好的咖啡使用起来非常方便，但同时也成为最糟糕的垃圾之一。

　　咖啡也可以根据咖啡因含量分为含咖啡因的咖啡和不含咖啡因的咖啡。

任务模块3 场景交际任务

场景1

A：嗨，埃文。很高兴见到你。

B：嗨，贾斯汀。我也很高兴再次见到你。昨天我去了时光咖啡馆。我对里面的饮品感到很困惑。你能告诉我咖啡饮品的分类吗？

A：在我看来，我们可以把咖啡饮品分为3类。它们是原味咖啡、经典咖啡和其他咖啡。但对于咖啡馆来说，它会以另一种方式对饮品进行分类。

B：先解释一下分类，好吗？

A：好的。让我们从原味咖啡开始。原味咖啡是不加任何添加物的咖啡。原味咖啡是不含任何添加剂的咖啡。我们通常把原味咖啡分为单品咖啡和拼配咖啡。它们都可以用任何设备冲煮而成。

B：单品咖啡？你能具体说说吗？

A：单品咖啡是指采用单一产区、单一采收季的咖啡豆制作的咖啡，如埃塞俄比亚的耶加雪菲咖啡、牙买加的蓝山咖啡、苏门答腊的曼特宁咖啡和中国云南的阿拉比卡咖啡。

B：那么拼配咖啡是不同咖啡的混合物，对吗？

A：我觉得是的。

B：那经典咖啡呢？

A：我认为经典咖啡通常以浓缩咖啡为基础，加了牛奶、奶油、巧克力酱或水等的热咖啡饮品。例如，卡布奇诺、拿铁、玛奇朵、馥芮白和欧蕾，这些咖啡都加了不同比例的浓缩咖啡和牛奶。摩卡咖啡是浓缩咖啡、巧克力酱和牛奶的经典组合。你可以使用浓缩咖啡、牛奶和稀奶油来制作布雷卫咖啡，还可以用不含牛奶的浓缩咖啡制作康宝蓝咖啡。美式咖啡、意大利超浓咖啡和廊构咖啡是含浓缩咖啡和水的饮品。

B：酷。那其他咖啡呢？

A：其他咖啡可以是冰的或热的，加入不同的东西，如酒精、酸奶、冰激凌、杏仁等。你可以通过添加不同的东西来创造不同的口味。

B：好像很有趣。你能给我冲一杯含酒精的咖啡试试吗？

A：没问题。你想要哪一种，威士忌、白兰地还是马提尼？

B：请给我马提尼。谢谢。

A：不客气。请稍等片刻。

场景2

A：贾斯汀，什么是精品咖啡？

B：这是一个术语，指的是具有独特风味的单一原味咖啡。

A：有什么标准来定义它吗？

B：有的。精品咖啡协会为精品咖啡设定了明确的评分标准。

A：我对精品咖啡的标准很好奇。

B：简单地说，如果SCA的杯测分数超过80分，咖啡就可以称为优质咖啡。

A：你可以跟我说说评分方面的细节吗？

B：可以的。根据标准，咖啡杯测（Coffee Cupping）分数高于80分的咖啡，称为精品咖啡。其中，得分为80～84.99分的咖啡被评为"非常好"，得分为85～89.99分的咖啡被评为"优秀"，而得分为90～100分的咖啡被评为"非常优秀"。

A：我明白了。精品咖啡通常来自哪里？

B：精品咖啡通常生长在"咖啡豆带"或"咖啡带"，包括南美洲、中美洲、亚洲和非洲。

A：世界上最贵的精品咖啡是什么？

B：我不确定。但据说巴拿马的瑰夏咖啡和埃塞俄比亚的耶加雪菲咖啡是两种最昂贵的精品咖啡。

A：我好想尝尝这些最好的精品咖啡。

B：我也是。

任务模块4　知识拓展

了解更多关于咖啡饮料的信息

在意大利，按传统，卡布奇诺在早上饮用，通常作为早餐的一部分，在上午11点前饮用。因为卡布奇诺是以牛奶为基础的，被认为味道太重了，不宜在更晚的时间饮用。卡布奇诺被广泛认为是咖啡和牛奶比例最协调的咖啡，这使得它成为以浓缩咖啡为基础的最受欢迎的咖啡饮品之一。这是卡布奇诺的经典配方。

1）你需要的设备

2个中杯

浓缩咖啡机

拉花缸

2）成分

16～20克精细咖啡粉。

130～150毫升冷鲜奶。

巧克力或肉桂粉可选。

3）可制成2杯。

4）温度：高温。

5）制作过程

第一步：用热水热杯或把杯子放在咖啡机上加热。

第二步：在每个杯子中冲泡一杯（25毫升）浓缩咖啡。

第三步：将牛奶加热20秒左右，打出细腻的奶泡。

第四步：将牛奶倒在每杯浓缩咖啡上，让奶泡浮在咖啡表面。为了在第一口咖啡中产生浓烈的咖啡味道，尝试在杯子周围保留一圈咖啡油脂。

第五步：根据喜好，可以用振动器或用筛子将一些巧克力粉或肉桂粉撒在泡沫上。

任务模块5 创意任务

这里有两个任务：一个是帮助人们更容易地理解精品咖啡的历史；另一个是帮助人们更多地了解不同咖啡饮品的配方。你和你的搭档可以选择其中任何一个任务。对于第一个任务，请写一份报告来展示你们的想法。对于第二个任务，请收集一些你们喜欢的咖啡饮品的信息。你们可以以下表为例。然后与他人分享你们的报告或配方。我们鼓励你们根据需要从图书馆或互联网获取更多信息。

浓缩咖啡饮品名称	成分和建议用量
Doppio 双份浓缩咖啡	2杯/50毫升浓咖啡
Caffè Latte 拿铁咖啡	1杯/25毫升浓缩咖啡和蒸牛奶（含约5毫米泡沫层）
Con Panna 康宝蓝咖啡	2杯/50毫升浓咖啡和1汤匙生奶油
Caffè Mocha 摩卡咖啡	4汤匙黑巧克力酱、蒸牛奶（含约1厘米泡沫层）、2杯/50毫升浓缩咖啡

任务模块6 任务执行能力考核

序号	考核细分项目	标准分数（分）	得分（分）
1	专业用语	10	
2	任务资料的基础训练	20	
3	场景任务1专项训练	20	
4	场景任务2专项训练	20	
5	知识拓展	10	
6	创意任务	20	

任务模块7　目标达成考核

评分考核项目	标准分数（分）	个人自评	小组互评	教师评分
理论知识	30			
技能训练	40			
职业素养	30			
总分（分）	100			
综合总分				
说明	综合总分=个人自评（占总分的20%）+小组互评（占总分的20%）+教师评分（占总分的60%）			

Project Four

Coffee Art

Task 1   Coffee Art

Part One   Task Introduction

1. Teaching Objectives

Knowledge Objectives	1. Understand the meaning and four basic categories of coffee art. 2. Experience the way that coffee art is used to communicate with others. 3. Learn about the positive impact of the development of coffee art on coffee beverage sales.
Skill Objectives	1. In the process of learning English, feel the way of thinking in English, and improve the logic, speculation and innovation of thinking. 2. Be able to use English to understand and express information, views and feelings more accurately, and conduct effective cross-cultural communication.
Competency Objectives	1. Cultivate the sense of collaboration and teamwork. 2. Broaden the international vision, respect the diversity of the world culture, and build a sense of community with a shared future for mankind. 3. Build confidence and set up a correct view of English learning. 4. Deepen the understanding of Chinese culture and enhance cultural self-confidence through cultural comparison.

2. Teaching Focuses

(1) Understand the meaning of coffee art and the positive impact of coffee art on communication and sales.

(2) Compare the differences between four basic categories of coffee art.

3. Teaching Difficulty

Effective intercultural communication in English.

 Part Two   Task Implementation

Task Module 1   Pre-tasks

Step 1   Group Cooperation

Work in pairs. Discuss the professional terms in the box and get familiar with them. You can look them up online with your smartphones if necessary.

> crema   microfoam   macrofoam   pattern   etching   stencil   3D art

Step 2   Panel Discussion

Work in pairs to discuss the following issues and exchange views.

1. Do you like coffee art? Are you willing to pay more for a coffee with coffee art on?

2. Do you know how to create the fantastic coffee art?

Task Module 2   Reading and Training

Step 1   Reading

Coffee Art

What is coffee art? Coffee art is an approach by creating a pattern or design with crema, microfoam, powdered food coloring and so on. It is not limited to latte coffee but also to other beverages like cappuccino or hot chocolate. The crema is a red or brown creamy emulsion on top of the espresso. The microfoam is the opposite of the macrofoam. Microfoam is shiny, slightly thickened, finely textured milk foam. Microfoam is also named velvet milk and microbubbles. Microfoam is often used for coffee art while macrofoam with visibly large bubbles usually used for cappuccinos. Powdered food coloring is commonly added into the steamed microfoam to create different colors of foam.

The coffee art is generally created by pouring the original or colored microfoam into an espresso. The key to creating different patterns is to maintain the crema as a contrast to the foam and to pour the foam in a perfect balance between height and speed. With the development of coffee art, some new forms appear. According to the different making methods, coffee art can be divided into four main types. The first one is free pouring which is just as mentioned above. The second one is etching. It is to use an etching tool, like cocktail stick, skewer or the sharp handle of a tablespoon, to create a pattern after pouring the microfoam. The third one is stencil design creating. It is to place a creative thick stencil on top of a poured coffee and sprinkle with flavored powder you like through the stencil. Then your design will be revealed after removing the stencil. The fourth one is the 3D art which is to mould and shape different patterns with the foam on top of poured coffees, the etching tools and food coloring as well as the stencil could be used if necessary. 3D art is especially popular for decorating drinks served for ladies and kids (like hot chocolate or babyccino).

Coffee art is hard to master. But it is possible for anyone to gain the technique with enough patience and practice. Once mastered, it could decorate a cup of coffee fantastically. Although coffee art has nothing to do with the quality of the coffee itself, the beautiful patterns and shapes indeed make the coffee more attractive. Many customers are willing to pay more for a coffee with coffee art on as they think that the more energy spent by a barista, the better a coffee will be, especially the one with the extra beautiful designs. Maybe it is one of the reasons why the World Coffee Art Championship could always attract bigger and bigger crowds.

 Word and Phrase Bank

New Words

pattern /'pæt(ə)n/ *n.*   图案，花样
crema /'kreɪmə/ *n.*   咖啡的油脂
latte /'lɑːteɪ; 'læteɪ/ *n.*   拿铁咖啡
emulsion /ɪ'mʌlʃn/ *n.*   乳状液
microfoam /maɪkrəʊ'fəʊm/ *n.*   微泡沫，微奶泡
textured /'tekstʃəd/ *adj.*   起纹理的
cappuccino /ˌkæpu'tʃiːnəʊ/ *n.*   卡布奇诺
espresso /e'spresəʊ/ *n.*   意式浓缩咖啡
etching /'etʃɪŋ/ *n.*   雕花（咖啡）
tablespoon /'teɪblspuːn/ *n.*   大调羹，大汤匙
stencil /'stensl/ *n.*   模板
sprinkle /'sprɪŋkl/ *v.*   洒，撒
fantastically /fæn'tæstɪkli/ *adv.*   奇特地

Phrases & Expressions

powdered food coloring   食用色素粉
contrast to   与……形成对比
be willing to   乐意，愿意
World Coffee Art Championship   世界拉花艺术大赛

Step 2　After-reading Training

I　Match the words on the left with their meanings on the right.

1. espresso	A. equal parts of espresso and hot milk topped with cinnamon and nutmeg and usually whipped cream
2. pattern	B. a spoon larger than a dessert spoon; used for serving
3. etching	C. a decorative or artistic work
4. microfoam	D. a person who makes and serves coffee in a coffee bar
5. crema	E. an example of properly steamed milk, considered ideal to pour coffee art
6. latte	F. a piece of paper, plastic, or metal which has a design cut out of it
7. barista	G. strong espresso coffee with a topping of frothed steamed milk
8. cappuccino	H. drawing an image into the latte with a thin, sharp instrument
9. stencil	I. a thin layer of foam at the top of a cup of espresso
10. tablespoon	J. black coffee brewed by forcing hot water under pressure through finely ground coffee beans

1. _____　2. _____　3. _____　4. _____　5. _____　6. _____　7. _____
8. _____　9. _____　10. _____

II　Read the statements and tick the correct boxes.

Statements	Right	Wrong	Not mentioned
Coffee art is the only approach by creating a pattern or design with crema, microfoam, and powdered food coloring.			
The crema is a red or brown creamy emulsion on top of the latte.			
Macrofoam is often used for coffee art.			
Coffee art is hard to master.			
Many customers are willing to pay more for coffee art.			

Answers to the Tasks

Task Module 3   Communication Tasks

Situation 1

Step 1   Role-playing

A: Christina, what do I need to create the coffee art?

B: What kinds of coffee art do you want to make?

A: The heart. I need to start with the basic designs. Please help me to do the practice.

B: OK. You'll need the espresso, a bottle of fresh freezing milk, a bucket of ice, a pouring pitcher, two medium cups and a damp cloth.

A: What is the ice used for?

B: Keeping the milk freezing.

A: Oh, I see. The freezing milk could help me to steam the milk for a longer time.

(Later)

B: Have you steamed the milk?

A: Yes. See? Smooth, slightly thick and shiny surface with no big bubbles at all.

B: Excellent. Do you remember the procedure?

A: Yes. Firstly, hold the cup with 25 mL espresso and put the handle facing my body. Then pour the milk into the middle of the crema from about 7 cm above the cup.

B: Wait. Before the pouring, tilt the cup about 45 degrees.

A: Oh, yes. After that I pour the milk, the crema will rise. Once the cup is two-thirds full, lower the pitcher closer to the cup. A heart shape will begin to form and become larger and larger while I keep pouring down the milk.

B: Great. Then what will you do?

A: When the cup is almost full, lift the pitcher back up and draw a line through the middle of the circle. A heart shape will be done.

B: You're very clear about the points of each step. Now let's start the practice. Keep patient and stable.

A: OK.

 **Word and Phrase Bank**

New Words

freezing /'friːzɪŋ/ *adj.*　冰冻的

bucket /'bʌkɪt/ *n.*　桶

pitcher /'pɪtʃə(r)/ *n.*　壶，拉花缸

medium /'miːdiəm/ *adj.*　中等的

damp /dæmp/ *adj.*　潮湿的

steam /sti:m/ *v.* 蒸；用蒸汽处理

smooth /smu:ð/ *adj.* 光滑的

surface /'sɜːfɪs/ *n.* 表面

bubble /'bʌb(ə)l/ *n.* 气泡，泡沫

tilt /tɪlt/ *v.* （使）倾斜

Step 2　Consolidation Training

Ⅰ　Match the words on the left with their meanings on the right.

1. freezing	A. cook something by letting steam pass over it
2. bucket	B. the outer boundary of an artifact or a material layer constituting or resembling such a boundary
3. pitcher	C. the withdrawal of heat to change something from a liquid to a solid
4. medium	D. slightly wet
5. damp	E. to incline or bend from a vertical position
6. steam	F. a round metal or plastic container with a handle attached to its sides
7. smooth	G. a hollow globule of gas
8. surface	H. describe something that is average in degree or amount, or approximately halfway along a scale between two extremes
9. bubble	I. a cylindrical container with a handle and is used for holding and pouring liquids
10. tilt	J. having a surface free from roughness or bumps or ridges or irregularities

1. _____　2. _____　3. _____　4. _____　5. _____　6. _____　7. _____

8. _____　9. _____　10. _____

Ⅱ　Decide whether the following statements are true (T) or false (F).

(　　) 1. To create the coffee art of heart, we only need the espresso and a bottle of fresh freezing milk.

(　　) 2. The heart is not the basic design of coffee art.

(　　) 3. The ice is used for keeping the milk freezing.

(　　) 4. The freezing milk could help to steam the milk for a shorter time.

(　　) 5. When the cup is almost full, remember to lift the pitcher back up.

Answers to the Tasks

Situation 2

Step 1   Role-playing

A: Good morning, Christina. What are you doing with these thick cards?

B: Good morning, Kevin. I'm making some stencils.

A: What are they used for?

B: I want to do some creative designs for my coffee art. Let me show you how to use them.

A: Can't wait. Is it a pagoda? Can you show me this one first?

B: Yes. Here is a cup of cappuccino with a rim of crema around the cup. Now you put the stencil on top of the cup. Make sure the card won't get wet.

A: OK. Then what should I do?

B: Now sprinkle the chocolate powder over the stencil.

A: Is it OK now?

B: No. More chocolate powder to keep the shape sharp. Now remove the stencil.

A: Perfect. It's simple and easy to replicate.

B: Yes. Here is the golden milk foam.

A: How could you make the milk foam golden?

B: With the food color powder.

A: I see.

B: Now I use the handle end of the spoon to dip some golden milk foam and then add a sun here. Etch a river in front of the pagoda with the toothpick.

A: Wow, it's really beautiful. I love it.

B: I'm glad to hear that. You can also design anything you like with your creative talent. No limits at all.

A: Can I have a try?

B: Certainly. Let's start from making a stencil with the card, the slicer and the cutting board.

 **Word and Phrase Bank**

New Words

pagoda /pə'gəʊdə/ *n.*   宝塔

sprinkle /'sprɪŋk(ə)l/ *v.*   撒

shape /ʃeɪp/ *n.*   形状

sharp /ʃɑːp/ *adj.*   锋利的；尖的

remove /rɪ'muːv/ *vt.*   移开；除去

replicate /'replɪkeɪt/ *v.*   重复，复制

dip /dɪp/ *v.*   浸，蘸

etch /etʃ/ *vi.*　蚀刻；雕刻

toothpick /'tu:θpɪk/ *n.*　牙签

slicer /'slaɪsər/ *n.*　切片刀

Phrases & Expressions

a rim of　边沿，外缘

the handle end of　手柄尾端

Step 2　Consolidation Training

I　Complete the following sentences with the words in the box. Change the form if necessary.

> rim　design　use　handle　show　try　cup　top　sprinkle　creative

1. What are they _____ for?

2. Can you _____ me this one first?

3. There is a _____ of crema around the cup.

4. _____ the chocolate powder over the stencil.

5. Here is a _____ of cappuccino.

6. I want to do some _____ designs for my coffee art.

7. You can _____ anything you like with your creative talent.

8. Now you put the stencil on _____ of the cup.

9. Use the _____ end of the spoon.

10. Can I have a _____?

II　Match the words on the left with their meanings on the right.

1. pagoda	A. having or made by a thin edge or sharp point; suitable for cutting or piercing
2. sprinkle	B. an Asian temple; usually a pyramidal tower with an upward curving roof
3. shape	C. take something away
4. sharp	D. do something in exactly the same way
5. remove	E. distribute loosely
6. replicate	F. the visual appearance of something or someone
7. dip	G. a small stick which you use to remove food from between your teeth
8. etch	H. a knife with a short, thin, sharp blade
9. toothpick	I. cut into the surface by means of acid or a sharp tool

Continuation table

10. slicer	J. immerse briefly into a liquid so as to wet, coat, or saturate

1. _____   2. _____   3. _____   4. _____   5. _____   6. _____
7. _____   8. _____   9. _____   10. _____

Answers to the Tasks

Task Module 4   Extensive Reading

Learn More about the 3D Coffee Art

3D coffee art becomes a new trend in the coffee world. 3D coffee art adds a totally different and creative dimension to coffee art. It becomes increasingly popular as the 3D designs are easier to master than the free pouring skills. A footprint, a bear or a rabbit could be simply sculptured and shaped with one or several tablespoons of microfoam. More animals, cartoon characters, flowers or other designs could be created once you have mastered the skills. What you need is one or two spoons, a cocktail stick, some additives (like hot chocolate, syrup, powdered food coloring and other materials with the flavor or color you like), the stencils (optional) and your rich creativity. You can make the patterns or shapes on top of a cup of espresso or milk with the microfoam. It is especially loved by coffee houses. Their baristas will decorate the coffee with different lovely and unique designs to make it more attractive. 3D coffee art has become a new fashion favored by young generations.

How to make a 3D rabbit on top of a cup of babyccino? You can create it by following steps:

Step one: Prepare a cup of babyccino with steamed milk. It's better to make the milk no hotter than 40 ℃ as it's for kids to drink.

Step two: Create some white microfoam in a pitcher. Make some red-colored foam with the food coloring in a cup. Don't forget the hot chocolate.

Step three: Add the white microfoam on the babyccino aiming for a 1 cm layer. Sprinkle a layer of chocolate powder on top of the foam. It forms a brown base of the rabbit sculpture.

Step four: Use two spoons to create a ball by scraping the foam from one spoon to the other. Then put the ball in the center of the chocolate base. The ball seems just like the head of a rabbit.

Step five: Module two ears. Add two small foam balls on the rabbit's head and shape the ears with the foam.

Step six: Dip some chocolate to etch the eyes, nose and beards of the rabbit. Add tongue to the rabbit with red foam. You can add any designs to make the rabbit more attractive.

Task Module 5　Creative Task

Coffee art could be made not only by a barista in a coffeehouse, but also by a coffee art hobbyist, even a common person might be able to successfully make it if with more patience and practice. Using the four main methods of coffee art, try to make different patterns and shapes with your partner. Remember to take photos to share with others and keep records of all your designs.

项目4

咖啡拉花艺术

任务1　咖啡拉花艺术

 第1部分　任务介绍

咖啡拉花艺术

1.教学目标

知识目标	1.理解咖啡拉花艺术的含义和4个基本类别。 2.体验咖啡拉花艺术是用来与他人交流的方式。 3.了解咖啡拉花艺术的发展对咖啡饮料销售的积极影响。
技能目标	1.在学习英语的过程中，感受英语的思维方式，提高思维的逻辑性、思辨性和创新性。 2.能够使用英语，更准确地理解和表达信息、观点和感受，能有效地进行跨文化交流。
素质目标	1.培养与人协作的精神和团队合作的意识。 2.开拓国际视野，尊重文化的多样性，树立人类命运共同体意识。 3.树立信心，树立正确的英语学习观。 4.通过文化比较，加深对中华文化的理解，增强文化自信。

2.教学重点

（1）理解咖啡拉花艺术的意义及其对传播和销售的积极影响。

（2）比较咖啡拉花艺术4个基本类别之间的差异。

3.教学难点

能用英语进行跨文化交际与沟通。

第2部分 任务实施

任务模块1 前置任务

步骤1 小组合作

两人一组，讨论下列方框中的专业用语并加以熟悉。如果需要，你可以使用智能手机在线查找。

> crema   microfoam   macrofoam   pattern   etching   stencil   3D art

步骤2 小组讨论

两人一组，讨论以下问题并交流观点。

1. 你喜欢咖啡拉花艺术吗？你愿意花更多的钱买一杯有咖啡拉花艺术的咖啡吗？
2. 你知道如何创造美妙的咖啡拉花艺术吗？

任务模块2 阅读与训练

咖啡拉花艺术

什么是咖啡拉花艺术？咖啡拉花艺术是一种用咖啡油脂、微奶泡、食用色素粉等创造图案或设计的方法。它不局限于拿铁咖啡，还包括其他饮品，如卡布奇诺咖啡或热巧克力。咖啡油脂是浮在浓缩咖啡表层的一种红色或棕色的奶油状液体。微奶泡与大奶泡相反。微奶泡是有光泽的、稍厚的、纹理细腻的牛奶泡沫。微奶泡也被称为天鹅绒牛奶和微气泡。微奶泡通常用于咖啡拉花艺术，而带有明显大气泡的大奶泡通常用于卡布奇诺。食用色素粉通常添加到蒸汽微奶泡中，以产生不同颜色的泡沫。

咖啡拉花艺术通常将原色或彩色的微奶泡倒入浓缩咖啡中。创造不同图案的关键是保持油脂与奶泡形成对比，并在倒入奶泡时实现高度和速度之间的完美平衡。随着咖啡拉花艺术的发展，一些新的形式出现了。根据制作方法的不同，咖啡拉花艺术可以分为4种主要类型。第一种是直接倒入成型法。第二种是雕花法。它是用雕花工具，如鸡尾酒棒、串或汤匙的锋利手柄，在倒入微奶泡后创建图案。第三种是模板设计制作法。这是将一个有创意的厚模板放在倒好的咖啡上，然后在模板上撒上你喜欢的调味粉。拿开模板后，你的设计就出现了。第四种是3D拉花艺术法。用倒在咖啡最上面的奶泡塑造不同的造型和图案，必要时可以使用雕花工具和食用色素以及模板。3D艺术在装饰女士们和孩子们喝的饮品（如热巧克力或贝比奇诺咖啡）时尤其流行。

咖啡拉花艺术很难掌握。但对任何人来说，只要有足够的耐心和练习，都有可能掌握这项技术。一旦掌握了这项技术，它就能把一杯咖啡装点得漂漂亮亮的。虽然咖啡拉花艺术与咖啡本身的质量无关，但美丽的图案和造型会使咖啡更具吸引力。许多顾客愿意花更多的钱买一杯带拉花艺术的咖啡，因为他们认为咖啡师花费的精力越多，咖啡就越好，尤其是设计特别漂亮的咖啡。也许这就是世界咖啡拉花艺术大赛总是能吸引越来

越多的观众的原因之一。

任务模块3　场景交际任务

场景1

A：克里斯蒂娜，我需要什么来创作咖啡拉花？

B：你想做什么样的咖啡拉花？

A：爱心。我需要从最基本的设计做起。请帮我做练习。

B：好。你需要意式浓缩咖啡、一瓶新鲜冷冻牛奶、一桶冰、一个拉花缸、两个中等大小的杯子和一块湿布。

A：冰是用来做什么的？

B：冷冻牛奶的。

A：哦，我明白了。冷冻牛奶可以让我有更长的时间加热牛奶。

（稍后）

B：牛奶你加热了吗？

A：加热了。看，表面光滑、稍厚、有光泽，完全没有大气泡。

B：非常好。你还记得操作流程吗？

A：记得。首先，握好装有25毫升浓缩咖啡的杯子，杯子把手朝向我的身体。然后从杯子上方约7厘米处将牛奶从咖啡油脂中间倒进去。

B：等等。在倒之前，将杯子倾斜45°。

A：哦，对。我倒入牛奶后，油脂会升上来。杯满三分之二后，将拉花缸放低，靠近杯缘。当我不断倒入牛奶时，心形开始形成并变得越来越大。

B：非常好。之后做什么？

A：当杯子快满时，提起拉花缸，在心形中间画一条线，心形就做好了。

B：你非常清楚每一步的要点。现在让我们开始练习。保持耐心和手稳。

A：好的。

场景2

A：早上好，克里斯蒂娜。你在用这些厚厚的卡片做什么？

B：早上好，凯文。我在做模板。

A：用来做什么？

B：我想为我的咖啡拉花艺术做一些创意设计。让我给你演示如何使用它们。

A：我都等不及想看了。这是一座宝塔吗？你能先给我看看这个吗？

B：是宝塔。这是一杯卡布奇诺，杯口周围有一圈咖啡油脂。现在你把模板放在杯子上面。要确保卡片不会被弄湿。

A：好的。下一步我该怎么做？

B：现在把巧克力粉撒在模板上。

A：现在可以了吗？

B：不够。多撒些巧克力粉保持形状凸显出来。现在移开模板。

A：完美。这个做法简单且容易复制。

B：是的。这是金色的奶泡。

A：你怎么把奶泡变成金黄色的？

B：加食用色素粉。

A：我明白了。

B：现在我用勺子的柄尖蘸点金色奶泡，在这里加个太阳。用牙签在宝塔前雕刻出一条河。

A：哇，太漂亮了。我好喜欢。

B：听到你喜欢我很高兴。你也可以用你的创造力设计任何你喜欢的东西。没有任何限制的。

A：我可以试试吗？

B：当然可以。让我们从用卡片、切片刀和砧板制作模板开始吧。

任务模块4　知识拓展

了解更多关于3D咖啡拉花艺术的信息

3D咖啡拉花艺术成为咖啡世界的一种新趋势。3D咖啡拉花艺术为咖啡艺术增加了一个完全不同的创意维度。3D咖啡拉花技术比直接倒入成型的拉花技术更容易掌握，它变得越来越流行。一个脚印、一只熊或一只兔子可以简单地用一勺或几勺微奶泡雕刻成型。一旦你掌握了这项技能，就可以创造更多的动物、卡通人物、花卉或其他设计。你需要的是一个或两个勺子、一根鸡尾酒棒、一些添加剂（如热巧克力、糖浆、食用色素粉和其他带有你喜欢的口味或颜色的材料）、模板（可选项）和你丰富的创造力。你可以在一杯意式浓缩咖啡或牛奶上用微奶泡制作图案或造型。咖啡馆尤其喜欢这样做。他们的咖啡师会用不同可爱的、独特的设计装饰咖啡，让咖啡更有吸引力。3D咖啡拉花艺术已成为年轻一代喜爱的新时尚。

如何在一杯贝比奇诺饮品上制作一只3D兔子？你可以按照以下步骤来制作。

第一步：用加热的牛奶准备一杯贝比奇诺，牛奶温度最好不超过40 ℃，因为这是孩子们喝的。

第二步：在拉花缸里制作一些白色的微奶泡。在一个杯子里用食用色素做一些红色奶泡。别忘了准备热巧克力。

第三步：将白色微奶泡添加到贝比奇诺上，厚度为1厘米。在微奶泡上撒一层巧克力粉，形成小兔造型的棕色底座。

第四步：用两个勺子将奶泡从一个勺子刮到另一个勺子上，形成一个球。然后将球放在巧克力底座的中心。这个球看起来就像兔子的头。

第五步：做两只耳朵的造型。在兔子的头上放两个小奶泡球，用奶泡做出耳朵的

造型。

第六步：蘸些巧克力，雕刻兔子的眼睛、鼻子和胡须。用红色泡沫为兔子添加舌头。你可以添加任何设计来使兔子更具吸引力。

任务模块5　创意任务

咖啡拉花艺术不仅可以由咖啡馆的咖啡师制作，也可以由咖啡艺术爱好者制作，如果有更多的耐心和练习，即使是普通人也可能成功制作。使用咖啡拉花的4种主要方法，尝试与你的伙伴一起制作不同的图案和形状。记住拍照与他人分享，并保留所有设计的记录。

任务模块6　任务执行能力考核

序号	考核细分项目	标准分数（分）	得分（分）
1	专业用语	10	
2	任务资料的基础训练	20	
3	场景任务1专项训练	20	
4	场景任务2专项训练	20	
5	知识拓展	10	
6	创意任务	20	

任务模块7　目标达成考核

评分考核项目	标准分数（分）	个人自评	小组互评	教师评分
理论知识	30			
技能训练	40			
职业素养	30			
总分（分）	100			
综合总分				
说明	综合总分=个人自评（占总分的20%）+小组互评（占总分的20%）+教师评分（占总分的60%）			

Project Five

Café Culture

Task 1　Café Culture

Part One　Task Introduction

1. Teaching Objectives

Knowledge Objectives	1. Learn about the appearing and development of café culture. 2. Correctly understand the role of café culture in the communication of culture and art.
Skill Objectives	1. How to use café culture to communicate Chinese culture. 2. In the process of learning English, feel the way of thinking in English, and improve the logic, speculation and innovation of thinking. 3. Be able to use English to understand and express information, views and feelings more accurately, and conduct effective cross-cultural communication.
Competency Objectives	1. Cultivate the sense of collaboration and teamwork. 2. Broaden the international vision, respect the diversity of the world culture, and build a sense of community with a shared future for mankind. 3. Build confidence and set up a correct view of English learning. 4. Deepen the understanding of Chinese culture and enhance cultural self-confidence through cultural comparison.

2. Teaching Focuses

(1) Learn about the appearing and development of café culture.

(2) Correctly understand the role of café culture in the communication of culture and art.

(3) How to use café culture to communicate Chinese culture.

3. Teaching Difficulty

Effective intercultural communication in English.

Part Two　Task Implementation

Task Module 1　Pre-tasks

Step 1　Group Cooperation

Work in pairs. Discuss the professional terms in the box and get familiar with them. You can look them up online with your smartphones if necessary.

> café coffeehouse non-caffeinated beverage Balzac French

Step 2　Panel Discussion

Work in pairs to discuss the following issues and exchange views.

1. How do you address a professional person who works in a café?

2. Do you have a dream of owning your own café in the future? If you have this dream, what preparations would you make in advance?

Task Module 2　Reading and Training

Step 1　Reading

Café Culture

The word café comes from French. In coffee culture, a café is also called a coffeehouse or a coffee shop. A café mainly serves coffee of various types. Some may also provide tea and other non-caffeinated beverages. In Europe, coffeehouses also serve alcoholic drinks. Besides drinks, many foods like doughnuts, sandwiches, muffins or fruit may serve the clientele.

Cafés offering an open public space soon sprang up across the world. People gathered in popular meeting places to drink coffee, exchange information, listen to stories and music, play board games, and discuss news and politics. Some people say that cafés are a paradise for artists and intellectuals as caffeine released here inspires a lot of imagination and innovative ideas. It's said that Balzac, the French novelist, would drink about 40 cups of coffee a day while writing in a café. He can't even write without coffee. Many believe that without coffee, there would be no *La Comédie Humaine*.

From a cultural standpoint, coffeehouses largely serve as important centers of social interaction. With the development of the times, no matter the scale of the coffeehouse is large or small, computers and Internet access help it become a contemporary-styled, youthful and modern place. For many people, sitting in a café with a cup of delicious coffee made by a skilled barista is just a normal part of improving the quality of everyday life, while for others it

is still a new, exciting and fascinating life experience. And rapidly the concept "Quality costs more." has been accepted by more and more people who love coffee.

 **Word and Phrase Bank**

New Words

café /kæ'fe/ *n.* 咖啡馆，咖啡屋，小餐馆

culture /'kʌltʃə(r)/ *n.* 栽培，文化，教养

French /frentʃ/ *n.* 法语

serve /sɜːv/ *v.* 提供，服务

various /'veəriəs/ *adj.* 各种各样的

alcoholic /ˌælkə'hɒlɪk/ *adj.* 酒精的，含酒精的，酗酒的

doughnut /'dəʊnʌt/ *n.* 甜甜圈

muffin /'mʌfɪn/ *n.* 小松糕，松饼

clientele /ˌkliːən'tel/ *n.* 顾客群

gather /'ɡæðə/ *v.* 聚集

paradise /'pærədaɪs/ *n.* 天堂

intellectual /ˌɪntə'lektʃuəl/ *n.* 知识分子

innovative /'ɪnəveɪtɪv/ *adj.* 新颖的，革新的

novelist /'nɒvəlɪst/ *n.* 小说家

standpoint /'stændpɔɪnt/ *n.* 立场，观点

scale /skeɪl/ *n.* 规模

youthful /'juːθfl/ *adj.* 年轻的

accept /ək'sept/ *vt.* 接受，同意

Phrases & Expressions

non-caffeinated beverages 无咖啡因饮料

an open public space 开放的公共空间

spring up 出现，兴起

across the world 世界各地

social interaction 社会交往

La Comédie Humaine 《人间喜剧》

contemporary-styled 现代风格的

Step 2 After-reading Training

Ⅰ Answer the questions.

1. Where did the word café come from?

_____

2. What kind of services does a café offer?

_____

3. What's the function of coffeehouses from a cultural standpoint?

_____

4. Why does the author say that cafés are a paradise for artists and intellectuals?

_____

5. What is the new concept "Quality costs more"?

_____

Ⅱ Read the statements and tick the correct boxes.

Statements	Right	Wrong	Not mentioned
The word café comes from French.			
In Europe, coffeehouses do not serve alcoholic drinks.			
It's said that Balzac, the French novelist, would drink about 50 cups of coffee a day while writing in a café.			
From a cultural standpoint, coffeehouses largely serve as important centers of social interaction.			
In coffee culture, a café is also called a coffeehouse or a coffee shop.			

Task Module 3   Communication Tasks

Answers to the Tasks

Situation 1

Step 1   Role-playing

A: Austin, how about your coffee training?

B: It's very funny. You know a café is a coffeehouse. But in fact it also has other meanings.

A: What are they?

B: The term café generally refers to a diner, British caff, greasy spoon, teahouse or other casual eating and drinking place.

A: What is British caff?

B: It's a small restaurant serving light and basic meals.

A: What is greasy spoon then?

B: It's a small and inexpensive restaurant that serves mostly fried food.

A: OK. What else did you learn?

B: Do you know what a coffee bean is?

A: It's a kind of bean, just like the soybean, isn't it?

B: No. They're not similar. A coffee bean is the pip inside the red or purple fruit. They are usually called cherry.

A: How can a cherry become a coffee bean?

B: When the cherry is ripe, people will process cherry to remove its skin. Then roast it before grind and brew it. A barista will make a good-quality coffee with the coffee powder for the consumers.

A: There is a specialty café opened at the crossroads. I smelled the strong aroma of coffee as I passed it.

B: So did I. And the café is named "Touch Your Heart".

 Word and Phrase Bank

New Words

casual /'kæʒuəl/ *adj.* 非正式的
soybean /'sɒɪbiːn/ *n.* 大豆，黄豆
pip /pɪp/ *n.* 果核
cherry /'tʃeri/ *n.* 樱桃
ripe /raɪp/ *adj.* 成熟的，熟的
roast /rəʊst/ *v.* 烤，烘焙
grind /graɪnd/ *v.* 磨，碾碎
barista /bə'riːstə/ *n.* 咖啡师
powder /'paʊdə(r)/ *n.* 粉末
crossroads /'krɒsˌrəʊdz/ *n.* 十字路口
aroma /ə'rəʊmə/ *n.* 浓香，香气

Phrases & Expressions

in fact 事实上
greasy spoon 小吃店
light and basic meal 便餐
fried food 油炸食品
coffee bean 咖啡豆
good-quality coffee 高品质咖啡

Step 2　Consolidation Training

I　Match the words on the left with their meanings on the right.

1. similar	A. reduce to small pieces or particles by pounding or abrading
2. grind	B. having the same or similar characteristics
3. aroma	C. a distinctive odor that is pleasant
4. pip	D. a piece of cutlery with a shallow bowl-shaped container and a handle
5. spoon	E. a small hard seed found in some fruits

1. _____　2. _____　3. _____　4. _____　5. _____

II　Decide whether the following statements are true (T) or false (F).

(　　) 1. A coffee bean is the pip inside the red or purple fruit usually regarded as a cherry.

(　　) 2. When the cherry is ripe, people will keep its skin. Then roast it before grind and brew it.

(　　) 3. British caff is a small restaurant serving light meals.

(　　) 4. A barista could not make a good-quality coffee for customers.

(　　) 5. There is a name for the specialty café referred at the end of the dialogue.

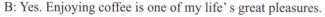

Situation 2

Answers to the Tasks

Step 1　Role-playing

A: Doris, it seems that you like coffee very much.

B: Yes. Enjoying coffee is one of my life's great pleasures.

A: Do you often go to the specialty cafés?

B: Yes. They are one of my favorite places.

A: I once thought that a specialty café is just for coffee connoisseurs.

B: It used to be. But nowadays, anyone could visit it and experience an array of varieties, roasts and styles of coffees.

A: In recent years an increasing number of specialty cafés open around us.

B: Indeed. I'm especially fond of staying in a café to read, write or pass the time. For others, it's a perfect place to socialize.

A: Since the café has supplied free Wi-Fi, many people especially the young frequent cafés more than ever before.

B: How often do you go to a café?

A: I usually go there at least twice a week.

B: How about going together this Saturday?

A: I'd love to.

 Word and Phrase Bank

New Words

pleasure /'pleʒə(r)/ *n.* 愉快, 高兴

connoisseur /ˌkɒnɪ'sɜ:/ *n.* 鉴赏家

array /ə'reɪ/ *n.* 大量

indeed /ɪn'di:d/ *adv.* 真正地，的确

socialize /'səʊʃəlaɪz/ *v.* 交往，交际

supply /sə'plaɪ/ *v.* 提供

frequent /'fri:kwənt/ *v.* 常去

Phrases & Expressions

coffee connoisseur 咖啡鉴赏家

an array of 大量

in recent years 最近几年

an increasing number of 越来越多

be fond of 喜欢

pass the time 消磨时间

twice a week 一周两次

Step 2 Consolidation Training

I Complete the following sentences with the words in the box. Change the form if necessary.

supply   around   seem   great   stay

1. Enjoying coffee is one of my life's _____ pleasures.

2. In recent years an increasing number of specialty cafés open _____ us.

3. I'm especially fond of _____ in a café to read, write or pass the time.

4. It _____ that you like coffee very much.

5. Since the café _____ free Wi-Fi, many people especially the young frequent cafés more than ever before.

II   There are ten English words and terms numbered 1 through 10 in the left column. You should match them with their Chinese equivalents marked A through J in the right column and write down the corresponding letter in the right blank.

1. an array of	A. 一周两次
2. twice a week	B. 多于
3. coffee connoisseur	C. 大量
4. more than	D. 喜欢
5. be fond of	E. 咖啡鉴赏家
6. in recent years	F. 的确
7. supply	G. 最近几年
8. pleasure	H. 供应
9. frequent	I. 愉快
10. indeed	J. 常去

1. _____ 2. _____ 3. _____ 4. _____ 5. _____ 6. _____
7. _____ 8. _____ 9. _____ 10. _____

Task Module 4   Extensive Reading

Answers to the Tasks

Learn More about Café Culture

With coffee consumption as the core, café culture is the carrier of inheriting a series of traditional cultures, cultural communication and social behaviors, and is regarded as the lubricant of social economy. People trace the culture being related to coffee and cafés back to 16th-century Turkey. Coffeehouses were considered not only as the social hubs but also as the artistic, intellectual, political and commercial centers. For example, the café named Les Deux Magots in Paris was once connected with the intellectuals Jean-Paul Sartre and Simone de Beauvoir. Now it is a very popular tourist attraction.

Elements of modern coffeehouses include inviting decor, alternative brewing techniques and slow-paced but exquisite food service. Many coffee shops offer access to free Wi-Fi for clientele, encouraging business, parties, study or other personal affairs at these locations. During coffee break, many employees from different work industries will gather here for a snack or short downtime, just as their routine social behavior. In order to attract more customers, many cafés will hire baristas with expert knowledge to advise customers on how to make best choices of their coffee.

Though café culture differs by places, coffee has played a large role in history and literature of these areas. And as a central theme, cafés culture has been considerably referenced in poetry, fiction and regional history as well as in comics, television and film.

Task Module 5　Creative Task

In order to pursue the dream of opening a café in the future, a lot of work including the preparation in technology, ideology, capital and management, needs to be made in advance. Work in a group. Try to discuss what you would like to prepare for the dream. You are encouraged to collect more information from the library or through the internet if necessary.

项目 5

咖啡馆文化

任务1　咖啡馆文化

第1部分　任务介绍

咖啡馆文化

1. 教学目标

知识目标	1. 了解咖啡馆文化的形成与发展。 2. 正确认识咖啡馆文化在文化艺术传播中的作用。
技能目标	1. 学习如何利用咖啡馆文化传播中国文化。 2. 在学习英语过程中，感受英语思维方式，提升自身思维的逻辑性、思辨性和创新性。 3. 能够运用英语，比较准确地理解和表达信息、观点、情感，能有效地进行跨文化交际与沟通。
素质目标	1. 培养与人协作的精神和团队合作的意识。 2. 拓宽国际视野，尊重文化的多样性，树立人类命运共同体意识。 3. 树立自信，树立正确的英语学习观。 4. 通过文化比较，加深对中国文化的理解，增强文化自信。

2. 教学重点

（1）了解咖啡馆文化的形成与发展。

（2）正确认识咖啡馆文化在文化艺术传播中的作用。

（3）如何利用咖啡馆文化传播中国文化。

3. 教学难点

能用英语进行有效的跨文化交际与沟通。

第2部分　任务实施

任务模块1　前置任务

步骤1　小组合作

两人一组，讨论下列方框中的专业用语并加以熟悉。如果需要，你可以使用智能手机在线查找。

café　coffeehouse　non-caffeinated　beverage　Balzac　French

步骤2　小组讨论

两人一组，讨论以下问题并交流观点。

1. 你怎么称呼在咖啡馆工作的专业人士？

2. 你有梦想在未来拥有自己的咖啡馆吗？如果你有这个梦想，你会提前做什么准备？

任务模块2　阅读与训练

咖啡馆文化

咖啡馆一词来自法语。在咖啡文化中，咖啡馆也被称为咖啡屋或咖啡店。咖啡馆主要供应各种类型的咖啡。有些咖啡馆还可以提供茶和其他不含咖啡因的饮品。在欧洲，咖啡馆也供应酒精饮料。除了饮品，许多食物如甜甜圈、三明治、松饼或水果也可能为顾客提供。

提供开放公共空间的咖啡馆很快在世界各地兴起。人们在这些广受欢迎的聚会场所，喝咖啡、交流信息、听故事、听音乐、玩棋盘游戏、讨论新闻和政治。有人说，咖啡馆是艺术家和知识分子的天堂，因为这里释放的咖啡因激发了很多想象力和创意。据说，法国小说家巴尔扎克在咖啡馆写作时每天大约会喝40杯咖啡。没有咖啡，他甚至不能写作。许多人认为，没有咖啡，就没有《人间喜剧》。

从文化角度来看，咖啡馆在很大程度上是社会交往的重要中心。随着时代的发展，无论咖啡馆规模大小，电脑和互联网的接入都使其成为一个具现代风格、年轻化和现代化的场所。对许多人来说，坐在咖啡馆里喝一杯由熟练咖啡师制作的美味咖啡只是提高日常生活质量的正常部分，而对于其他人来说，这仍然是一种全新的、令人兴奋的、迷人的生活体验。而且很快，"品质更有价值"的概念被越来越多热爱咖啡的人所接受。

任务模块3　场景交际任务

场景1

A：奥斯汀，你的咖啡培训怎样了？

B：很有意思。你知道咖啡馆就是咖啡屋。但事实上，它也有其他含义。

A：什么含义？

B："咖啡馆"一词通常是指餐厅、英国咖啡馆、小吃店、茶馆或其他休闲餐饮场所。

A：什么是英国咖啡馆？

B：就是供应便餐的小餐馆。

A：那小吃店是什么？

B：小而便宜的、主营油炸食品的小餐馆。

A：好。你还学到了什么？

B：你知道咖啡豆是什么吗？

A：一种豆子，就像大豆一样，不是吗？

B：不，它们并不像。咖啡豆是红色或紫色果实里的果核。它们常被称为樱桃。

A：樱桃怎么能变成咖啡豆？

B：当樱桃成熟时，会对它进行加工，去掉表皮。然后进行烘焙、研磨和冲煮。咖啡师会用咖啡粉为消费者制作高品质的咖啡。

A：十字路口那里开了一家特色咖啡馆。当我经过时，我能闻到咖啡浓浓的香味。

B：我也是。那个咖啡馆叫"触动你的心"。

场景2

A：多丽丝，你似乎很喜欢喝咖啡。

B：是的。享受咖啡是我生活中最大的乐趣之一。

A：你经常去特色咖啡馆吗？

B：是啊。那是我最喜欢去的地方之一。

A：我曾经以为特色咖啡馆只适合咖啡鉴赏家去。

B：过去是这样的。但是现在，任何人都可以去咖啡馆，去体验不同咖啡的品种、烘焙和风格。

A：近年来，我们周围开了越来越多的特色咖啡馆。

B：的确如此。我特别喜欢待在咖啡馆里读书、写作或打发时间。对其他人来说，这是一个完美的社交场所。

A：自从咖啡馆提供免费Wi-Fi以来，许多人尤其是年轻人比以往任何时候都更常去咖啡馆。

B：你多久去一次咖啡馆？

A：我通常每周至少去两次。

B：这个星期六一起去怎么样？

A：好啊。

任务模块4　知识拓展

了解更多关于咖啡馆文化的信息

咖啡馆文化以咖啡消费为核心，是传承一系列传统文化、文化传播和社会行为的载体，被视为社会经济润滑剂。人们将与咖啡和咖啡馆有关的文化追溯到16世纪的土耳其。咖啡馆不仅被视为社会中心，还被视为艺术、知识、政治和商业中心。例如，巴黎名为Les Deux Magots的咖啡馆曾与知识分子Jean-Paul Sartre和Simone de Beauvoir联系在一起。现在，它是一个非常受欢迎的旅游景点。

现代咖啡馆的元素包括有吸引力的装饰、独特的冲煮技术和节奏慢但很讲究的餐饮服务。许多咖啡馆为顾客提供免费Wi-Fi，鼓励他们来这里开展商务活动、聚会、学习或其他个人事务。在喝咖啡休息期间，许多来自不同行业的员工会聚集在这里吃点心或短暂休息，就像他们的日常社交一样。为了吸引更多的顾客，许多咖啡馆会雇用有专业知识的咖啡师，就如何选择最好的咖啡向顾客提供建议。

虽然咖啡馆文化因地而异，但咖啡在这些地区的历史和文学中发挥了重要作用。作为一个中心主题，咖啡馆文化在诗歌、小说、地区历史以及漫画、电视和电影中都被大量引用。

任务模块5　创意任务

为了追求未来开设咖啡馆的梦想，需要提前做好技术、思想、资金和管理等方面的准备工作。小组合作，试着讨论一下你们想为这个梦想做些什么准备。我们鼓励你们根据需要从图书馆或互联网收集更多信息。

任务模块6　任务执行能力考核

序号	考核细分项目	标准分数（分）	得分（分）
1	专业用语	10	
2	任务资料的基础训练	20	
3	场景任务1专项训练	20	
4	场景任务2专项训练	20	
5	知识拓展	10	
6	创意任务	20	

任务模块7　目标达成考核

评分考核项目	标准分数（分）	个人自评	小组互评	教师评分
理论知识	30			
技能训练	40			
职业素养	30			
总分（分）	100			
综合总分				
说明	综合总分=个人自评（占总分的20%）+小组互评（占总分的20%）+教师评分（占总分的60%）			

参考文献

[1] （英）班克斯（Banks, M.），（英）麦费登（Mcfadden, C.），（英）埃克丁森（Atkinson, C.）. 咖啡圣经：从简单的咖啡豆到诱人的咖啡的专业指南[M].徐舒仪，译. 北京：机械工业出版社，2014.

[2] （澳）奥尔加·卡里耶（Olga Carryer）. 咖啡师圣经[M]. 潘苏悦，译. 北京：机械工业出版社，2017.

[3] 日本枻出版社编辑部.完全咖啡知识手册：升级版[M]. 张文慧，译. 北京：中国轻工业出版社，2022.